STAIRWAYS TO THE STARS

SKYWATCHING IN THREE GREAT ANCIENT CULTURES

Anthony Aveni

John Wiley & Sons, Inc.

New York • Chichester • Weinheim • Brisbane • Singapore • Toronto

Cassell Publishers Limited
Wellington House, 125 Strand
London WC2R OBB

Copyright © Anthony F. Aveni 1997

First published in Great Britain 1997
by arrangement with
John Wiley & Sons, Inc.,
605 Third Avenue
New York, NY 10158

British Library Cataloguing-in-Publication Data
A catalogue record for this book is available from the British Library

ISBN 0-304-34998-4

Printed and bound in Great Britain by Creative Print and Design Wales, Ebbw Vale

To Walter F. Fullam
for supporting
Colgate and Archaeoastronomy

CONTENTS

PREFACE

Albert Einstein once suggested that we should not wonder that other civilizations in the world had failed to form the synthesis necessary to give us our Western scientific view of nature—the miracle is that such a synthesis ever took place at all. The miraculous coming together Einstein spoke about is well known to anyone who studies science. The logic of geometry—a gift from the Greeks—fused with arithmetical skills derived from the Babylonian culture gave rise to a highly abstract notion of an idealized world that speaks to us through the language of mathematics. It is an elusive world veiled beneath phenomenal imperfections revealed through the senses. During the Renaissance, more than a thousand years after the demise of the classical world, the spirit of "meeting the test"—subjecting idealized hypotheses about the perceived world to the acid test of experimental inquiry in order to precisely quantify nature's behavior—sealed a new covenant between mind and matter. And while it may be true that never has there been another development quite like it in all of human history, Einstein may have failed to appreciate that science constitutes but one of a multitude of meaningful ways of probing and acquiring systematic knowledge about the world around us.

Stairways to the Stars selectively explores the diverse evidence concerning some of these other ways of knowing nature—the sky in particular—and compares them with our own. The principal barrier standing in our path will be our own addiction to the way modern science has taught us to view space and time. We tend to see little value in studying cultures that believed in a flat earth or an earth-centered cosmos with layers of heaven, and we have even less patience with ancestors who thought that events in the natural world repeated themselves in cycles. Yet in the history of the world no culture but our own—and then only very recently—has chosen the path of pursuing the ideology of an infinite space-time that frames colossal cataclysmic events, with humanity relegated to the role of insignificant bystander. "This is where science leads us," we say. "Can we help it that this is the true nature of the universe that we have unmasked with our quantitative experimentation?"

There is irony in the humbling course of events that followed science out of the scientific Renaissance, for by practicing science we have become accustomed to placing our worldview on an unreachable pedestal of progress, a plateau we view as unattainable by all the other cultures of the world, whom we tend to regard as less advanced than ourselves. Those "other," from Bronze Age to Babylon, have something to offer us, though; their brains were no less advanced than our own, their minds no less inquiring, their ways of categorizing knowledge no less systematic even if directed toward broader goals. Most importantly, the object of study for astronomer past and present—the sky—was very much the same. We need to get in touch with what we have forgotten lest we fall irretrievably deep into the chasm of cultural self-centeredness. Dare we really believe we are somehow more worthy, more special than all those who have passed before us in the great cosmic scheme? Can it really be true that only what *we* do is legitimate and what everyone else did was pretending?

Archaeoastronomy is the study of the practice of astronomy by ancient cultures of the world. It bases its evidence on an examination of unwritten as well as written remains, and serves as an interdisciplinary bridge connecting the established fields of astronomy, archaeology, culture history, and the history of astronomy. Archaeoastronomy has come a long way since the 1960s when it first came to be widely practiced. For example, field workers are long past the age of simply measuring and reporting the facts of astronomical building alignments in the archaeological record and calendrical correlations in the manuscripts in order to learn just how much "they" are (or are not) like "us." Today's archaeoastronomy is much more sophisticated and substantial, and far more interdisciplinary in scope, than it was twenty years ago. Researchers now write about comparative religion, art history, ethnology, ethnohistory, and the history of science. The study of astronomy has become a part of the study of human culture.

I designed *Stairways to the Stars* for readers interested in history and anthropology as well as science. I intend it to be a source of understanding how the art of skywatching developed in the Western tradition as well as in other major cultures of the world. To exemplify how archaeoastronomy has evolved over the past three decades, I have chosen to write about the remains of three well-documented, quite diverse culture areas that I believe best demonstrate the way we archaeoastronomers use unwritten evidence to understand the nature of astronomy and its role in the society that practiced it. The first, Stonehenge, represents a thoroughly preliterate culture in which the role of astronomy has proven to be quite controversial. Our second area, the Maya, emanates from a literate people known for their sophistication in calendrics and numeration. It also represents an area of the world where discoveries by archaeologists and script decipherers have unfolded at a staggering pace. Finally the third culture area, the Inca, represents a highly urbanized hierarchical society that developed an extensive empire. Their world has only recently begun to attract attention, thanks again to major archaeological findings.

To assist those who would like to be able to see the sky the way the ancients did, I have preceded the presentation of all the culture material with a lesson in simple naked-eye sky watching. For those who would reflect more deeply on the subject, a series of exercises, "Things to Think About," appears in Appendix A. Also, I have added a short description of archaeoastronomical field methods in Appendix B. Selected readings are recommended for readers who might care to pursue additional case studies. Finally, I have appended an annotated research bibliography for those who might wish to probe further the three cultures or to look into what we know about the astronomies of ancient cultures in other parts of the world.

My gratitude goes out to three very special people: Jacqueline D'Amore, Patricia Ryan, and Lorraine Aveni who assisted in the preparation of this manuscript and its illustrations, to Clive Ruggles for his advice, to Peter Tagtmeyer for his bibliographic search, to Jessica Deckard, who helped with proofreading, and to Ellen Walker, who did most of the drawings.

Thanks are also due Faith Hamlin, my agent, and Emily Loose and her staff at Wiley, especially Joanne Palmer, to David Golante, the eagle-eyed indexer, and to Jose Almaguer for the beautiful cover design.

Anthony Aveni

CHAPTER ONE

INTRODUCTION: A DIFFERENT SORT OF COSMOS

> The heavenly motions are nothing but a continuous song for several voices (perceived by the intellect, not by the ear); a music which, through discordant tensions, through sincopes and cadenzas ... progresses toward certain predesigned, quasi six-voiced clausulas, and thereby sets landmarks in the immeasurable flow of time.
>
> —Johannes Kepler,
> 1620 Dedication of *Ephemerides* to Lord Napier

The Lakota Sioux say that when the constellation of the Chief's Hand (the lower portion of our Orion) disappears from the sky, the earth will become infertile. They tell the story of a man who seeks to marry a woman from a village nearby. She will accept only the one who is able to recover her father's hand, which had been taken away by the sky gods as punishment for his refusal to make sacrifices to them. As the man travels from village to village on his quest, he is given gifts of special powers from the spirits. Eventually he defeats the supernaturals, acquires the hand, restores it to the chief, and marries the woman—who later gives birth to a son. For the Lakota sky watcher, the return of the hand to the sky signals the beginning of a new year to replace the old one that was interrupted by the chief's self-centeredness. The newborn son is proof of the continuation of the god-given power of fertility. In summer, just before Orion returns to the sky, the Lakotas conducted a ceremony of blood sacrifice. They listen to the divine speech manifests itself in the form of constellations, and they watch for celestial signals that will anticipate the stars' movements. As a reminder of their debt to the gods for fructifying mother earth, they tell and retell the story of the Chief's Hand and many other sky stories that give meaning to their lives.

All of this may sound like silly superstition to us, but suppose for a moment that you were not aware that we all live in an explosively expanding universe reaching out billions of light-years in all directions—a universe of distant rotating pinwheel isles made up of hundreds of billions of stars like the sun, each in its own orbit. Suppose, too, that you knew nothing of the vast dimensions of time that persist in this indifferent, unreachable sea of space—of either a beginning of the universe as we know it, a dozen billion years ago, or a possible end to it billions of years in the future. Imagine going out under a pristine, star-studded sky, far away from all the lights of the city, neither knowing nor caring about such incomprehensible spatial and temporal dimensions. My guess is that you would probably feel so close to the stars that you could reach out and touch them. Once you began to discern their behavior and saw that they present themselves as perfect role models—steady, dependable, predictable, and all functioning together like a well-ordered state or empire—you might actually think about talking to the bright lights that move about the sky and inquire into their animate wills in order to understand their personalities, their range of powers.

You might even plead to the forces of nature to intercede on your behalf:

> *Scorching Fire, warlike son of Heaven,*
> *Thou, the fiercest of thy brethren,*
> *Who like Moon and Sun decidest lawsuits—*
> *Judge thou my case, hand down thy verdict*[1]

reads an old Mesopotamian incantation. This is why all developing civilizations paid attention to the sky. The cyclic movement of the sun, moon, planets, and stars represented a kind of perfection mere mortals could strive after. What happens in the sky mirrors what happens in daily life. The regular occurrence of sunrise and moonset provided our ancestors with a concept of order, a stable pillar to which they could anchor their minds and souls.

In an age concerned with acquiring knowledge we never knew, we have perhaps paid too little attention to knowledge we have forgotten. Who knows the time the sun rose this morning or the current phase of the moon? We no longer need this sort of practical knowledge in our daily lives. Unlike our ancestors, we spend most of our time in a regulated climate with controlled lighting, totally detached from the natural environment. Technology has created an artificial backdrop against which we play out our existence. Any need we once had to watch carefully for celestial events has been lost; sky knowledge once held vital is now deemed useless. The artificial clocks by which we pace our daily activities have given us a distorted view of the dependence of real time on circumstances transpiring in the heavens.

Speculate we may, but we really cannot appreciate how much the minds of our predecessors were preoccupied with the cosmos. Heaven and nature touched every aspect of ancient culture, so it is no wonder we find sky stories woven into myth, religion, and astrology. So great was the ancients' reliance upon the sun and moon that they even deified them. For the Greeks, Apollo paraded in his daytime chariot across the sky. For the Aztecs, Tonatiuh's flint-knived tongue beckoned for the blood of sacrifice—human payment for the debt owed him for lighting up the world:

> The sun hath come to emerge. . . . But how will he go on his way? How will he spend the day? . . . They said unto him: Perform thy function! Work, O, our lord![2]

At regular intervals throughout the night before each feast day they offered him incense and they fasted. They cut their ear lobes, drew blood, and flicked it skyward with straws, a Spanish witness tells us.

Representations of eternal luminaries in the sky adorned ancient temples as objects of worship; they were symbolized in sculpture and painting. Our forebears followed their sky gods' movements attentively. By marking their appearance and disappearance with great care, they combined religious worship with practical knowledge. For example, the sun's return to a certain place on the horizon signaled when to plant the crops, when the river would overflow its banks, or when the monsoon season would arrive. The cycle of planting and harvesting crops was regulated by celestial events; important days of celebration and festivity were marked in a memorized celestial calendar, a symphony whose notes are the punctuated events of appearance and disappearance of celestial objects. Equipped with a knowledge of mathematics and a method for keeping permanent records, some cultures were able to refine and expand their knowledge of positional astronomy. After several generations, and with the advantage of a written record, they learned to predict particular celestial phenomena, such as eclipses, well in advance.

Understanding what happens in the sky is the basic prerequisite for appreciating other people's conception of the heavens. Here we begin the task of recovering lost knowledge. Our most important question is: What are the significant sky events that might have been watched by the ancients? Given no technological aids—no telescopes, no computers, no clocks—how can we determine the time and place of occurrence of such events, and how accurately can we observe them? How has the appearance of certain astronomical phenomena changed since the time these cultures developed? Finally, how can we retrieve astronomical information from an examination of the past's cultural remains—from the orientations of ancient pyramids, or from inscriptions carved on monuments or written down in texts?

Because ancient people generally believed that their spiritual and social lives were linked with the material world, they expended considerable effort paying tribute to celestial deities. We should not be surprised to find that in many instances, astronomical knowledge played a role in the design of their physical spaces—the cities and the sacred temples where they worshipped their gods. Stonehenge is perhaps the most famous example of an ancient structure believed to have served an astronomical function. Astronomer Gerald Hawkins's popular book *Stonehenge Decoded* (1966) rekindled the fires of archaeoastronomy and attracted a great deal of attention to this 5,000-year-old circular temple.

Anyone who approaches Stonehenge in a tour bus marvels at the stark sight of the 100-foot-wide ring of nearly three dozen 20-foot-tall gray stone slabs perched in the middle of southern England's gently rolling landscape. What was there 5,000 years ago—a great ceremonial accessway more than a mile long, communal roundhouses, outlying places of gathering and worship all built of thatch—has vanished.

Hawkins and other scholars (most of them trained in the sciences) who analyzed it in the 1960s saw the great monument as a precise observatory and computing device designed to predict eclipses. Were the scientists looking for an archaic version of their own technology? Their work raised many questions: Why would these people look at the sky? What useful knowledge might they gain from doing so? What advantage could they acquire to enable them to adapt to their living environment? Why would Bronze Age Britons erect massive stone architecture to align specifically with the sun and moon? Can such standing stones really be arranged to predict eclipses? (You really need to know how the sky works to answer this question!) And, most importantly, how do such theories resonate with what we actually know about these cultures based on archaeological evidence?

Of Stonehenge, the historians asked: How could a primitive culture accomplish great astronomical feats even before the Egyptians and Babylonians, given little other evidence that they were skilled mathematicians? Unfortunately, the word "primitive" wakes in us the image of a hunched hairy brute, club in hand, clad only in a loincloth. In fact, the modern archaeological records tell us these Britons were not so primitive at all. They were well organized, living in communities in which people performed specialized tasks. They were well dressed and well groomed, and by trading they managed to keep a bountiful hearth.

Hawkins's findings, if true, would turn history on its ear. But further studies showed that the precise astronomy he was looking for may not have been so precise after all, even if a concern with sky predictions was still a part of the archaic building plan. We are going to follow the sinuous course of development that has resulted in the Stonehenge model we have molded today. The controversy over the level of scientific prowess of our preliterate ancestors continues, but a few issues are resolvable, such as the major one: What tools do you need to make precise scientific predictions? I want to use the Stonehenge controversy as a way of taking up these concerns.

Since the 1960s, archaeoastronomy has moved away from the study of building alignments toward trying to understand the role of astronomy in ancient cultures in general—in other words toward the history of astronomy. One difference is that history usually deals with literate societies and focuses largely on analyses of traditional ancient scriptures such as hieroglyphs penned on papyri and symbols impressed on cuneiform tablets. Where the written evidence is sparse, archaeoastronomy has developed into a broader inquiry that relies more substantially upon the archaeological and iconographic record.

Though investigators were strongly attracted to the megalithic sites in Europe in the 1960s, by the 1970s they began to measure and map out astronomical building orientations in other parts of the world, particularly in the Americas. By the mid-1980s and 1990s, books dealing with the astronomies of ancient cultures of Italy, Mexico, and Peru had been published—a whole volume dealt with the Canary Islands alone. (See Appendix C for some of these references.) Though there is much chaff among the wheat in many of the treatises, more than enough new information and ideas remain to captivate the interest of those whose lifework is studying culture. In the past two decades many anthropologists and historians have become attracted to these new discoveries. They have become more seriously interested in just how important astronomy was and precisely where a people's knowledge of the sky fit into the picture.

Perhaps nowhere is the use of celestial knowledge more dramatic in its impact than in the world of the ancient Maya. Their rulers carefully tailored their pursuits, or so they tell us in their monumental inscriptions, after the comings and goings of particular celestial luminaries, Venus especially. After assessing the Bronze Age astronomy of Stonehenge, we turn to how and why the Maya's star-crossed destiny was rooted in celestial observation.

Archaeologist Sir Eric Thompson once suggested that to understand Maya astronomy one needed to get into the skin of the Maya astronomer. He meant that a knowledge of the history and culture of these Native American people was vital if we would ever hope to understand their celestial systems. Input from the archaeological discipline is important, too, because it represents a large part of the surviving record.

As an astronomer I became fascinated by the Maya because they seem to have been carried away with their celestial obsessions. The precision of the mathematical expression associated with their skywatching rivaled that of Babylonia and Greece and they managed all of it with a relatively low technology. They built skyscraper pyramids to penetrate the top of the jungle canopy that enveloped most of their culture for more than a thousand years. Another peculiarity (at least in our eyes) is that Maya astronomy was driven by astrology—using the stars to predict the future course of a ruling elite who believed they were the direct descendants of the gods. They even publicly offered their blood to the heavenly ancestors to reinforce their belief.

Remembering the lesson of Stonehenge, those of us trained in the modern sciences must be careful not to slant our view of our predecessors too much toward the present. And we cannot assume that the Maya were always looking for the same celestial events that matter to us. As we immerse ourselves in Maya cultural astronomy we will need to keep this in mind.

When their work of setting the stage for the creation of humans was finished, the twin ancestor gods, founders of the lineage of the Quiché Maya (who tell the story), rose into the midst of the sky; one became the sun, the other the moon. The 400 boys who assisted them followed, and they were turned into stars. Part of a larger cultural complex known as Mesoamerica, which stretched from today's United States–Mexico border to Honduras and Guatemala, the ancient Maya culture developed very different ideas about the cosmos than our own. Their cosmology was based on a broader kind of faith in nature than our secular world harbors. They believed that everyday life is intimately related to an animate natural world and that these two spheres are meant to function in harmony. The universe is a distinct whole, with all its parts intricately laced together, each aspect influencing the others. Nature and culture are one. Maya sky myths explained the unfolding of history, politics, and social relations, as well as ideas about creation and life after death. The Maya forged links between the sky and just about every component of human activity. And they celebrated this knowledge not only in their written books but also in their art, architecture, and sculpture. Recent popular works have praised the Maya for their precise predictions based on a knowledge of the sky. Are we witnessing a trend or is this a revelation that will stand the test of time? How can we establish exactly what Maya astronomers were watching? Can we really determine how they employed their knowledge? To investigate these questions in depth, we are going to zero in on one planet, one written document, and a handful of architectural and artistic expressions of Maya knowledge of a primary sky deity: Venus.

When we think of world capitals planned from the ground up, Brasília and Washington quickly come to mind. But the Inca nobles, who conquered all the territory and assimilated all the people from Ecuador to Chile in a lightning sweep just a few generations before the Spanish contact with the New World, outdid anything Western architects might have had in mind had they built Cuzco, the capital of the Inca empire in the highlands of southern Peru. Imaginary sight lines radiated like the spokes of a wheel reaching out from the Temple of the Ancestors, the navel of the Inca world, to the extremities of the empire and the universe. These invisible lines (they called them *ceques*, meaning "rays") divided all the universe into kin-related groups. Cuzco's great urban plan also tied together the sacredly held organization of land, water, mountains—even the stars. It was literally a map of the world graphed and plotted on top of the world itself, cast in a reference frame of religious worship based on sacrifice and offering to a living entity of which all people were a part.

We can be sure that astronomy was among the categories of precise knowledge encoded in the *ceque* system of Cuzco. The Inca had observatories with windows to catch the first and last rays of the sun, the half Inca-half Spanish chronicler Felipe Guaman Poma de Ayala wrote shortly after the fall of Cuzco. Another writer, Bernabe Cobo, detailed astronomical alignment schemes that appear to have made use of observation points connected to high mountains surrounding Cuzco. There particular *huacas* (sacred places) along the visible horizon were employed to mark the positions of important celestial objects at times of the year deemed worth noting. By direct sighting of sunrises and sunsets over particular *huacas*, the Inca preserved an accurate record of the most important dates of the seasonal year without needing to write it down.

Inca astronomy was highly socialized. And so, to understand where astronomy fits into the *ceque* system of the highly urbanized, hierarchically structured capital of Cuzco, we will need to sketch out the political, social, religious, and economic underpinnings of their astronomy. Only then can we entertain the question of how the Inca employed Cuzco's land-and-skyscape to devise a calendar and system of alignments that charted the course of the year. How did astronomy affect both agriculture and ancestor worship in the dominion of the Inca, and how was it adapted to transfer the ideology of the capital city to the remote hinterlands of the empire? Analyzing what the Spanish chroniclers have to say about astronomy (this means getting a sense of their sixteenth-century cultural and scientific biases, too!) will be seminal to the understanding of Inca astronomy, for unlike the Maya, no decipherable precontact written record survives.

Lastly, I want to address the important question: How is the history of our scientific astronomy different from those of the three areas—Stonehenge, Maya, and Inca? To ask it we need to raise another question: What happened in Greece and during the Renaissance that caused Indo-European cultures to take a turn away from the direct approach to nature taken by most other cultures of the world? *Stairways* concludes by highlighting the great turning points in the science of astronomy in our own Middle East-European past that serve as the foundation that sets apart the West from the rest regarding the perception of nature—Einstein's miracle.

There is a ring of familiarity in ancient skyways and modern science's quest for order and precision in nature, so beautifully encapsulated in Johannes Kepler's statement quoted at the beginning of this chapter—a familiarity diminished by the odd, subjective notion that nature is really out there waiting to engage in a dialog with us. My closing chapter, then, has dual goals: First, to link the study of the modern scientific discipline of astronomy to its roots in the past; and second, to highlight the similarities and differences between science as we know it and the astronomical pursuits of diverse cultures of the world. Only by establishing these foundations can we better appreciate our own contemporary astronomy. Only by understanding others can we hope to understand ourselves.

CHAPTER TWO

THE NAKED SKY

O Great ones, gods of the night . . .
O bow star and yoke star
O Pleiades, Orion and the Dragon
O Ursa Major, gout star, and the bison
Stand by, and then,
In the divination which I am making
In the lamb which I am offering
Put truth for me.
—Old Babylonian Prayer to the Sky Gods

Even modern astronomers are not used to watching the movements of the sun, moon, planets, and stars as we are about to describe them. Sure, they track Jupiter and Mars among the stars but the contemporary watchers of the skies look up, not ahead. They have no real interest in noting either where the planets enter the sky from the hazy horizon or where they leave us to pass back into the underworld. How long during the course of the year a planet is around before being swallowed by sunlight is unimportant, too. But people in other times and places fashioned their own cosmos out of quite different celestial perceptions.

In order to understand what ancient people thought about the world around them, we must begin by witnessing phenomena through their eyes. A knowledge of each particular culture is necessary, but learning what the sky contains and how each entity moves is also indispensable. Most important of all, we must remember that the ancient window to the stars was unencumbered by the technological devices of modern times such as the telescope, which, though it vastly extends our vision, also has altered our interpretation of what we see.

In its time, the kind of knowledge we are going to talk about in this chapter was important because of the power it bestowed upon those who possessed it. The ability to predict the future—to foresee the whereabouts of the luminaries who ruled the sky and the earth—gave prestige, respect, and authority to the adept who bore the gift. Strange but true: Whole cities, kingdoms, and empires were founded based on observations and interpretations of natural events that pass undetected under our noses and above our heads.

TROPIC AND TEMPERATE SKY

Whether ancient or modern, everything that happens above us appears to take place on the inner surface of a huge screen having roughly the shape of a hemisphere with the observer at the center. We can think of the *celestial sphere* as a sphere of arbitrary extent centered about an observer located at some position on earth. The situation as most of us would imagine it is pictured in Figure 2.1 with the size of the sphere vastly shrunken to fit the page. In Figures 2.2 and 2.3, two sets of views of the sky depict the nightly paths of celestial bodies seen just above the horizon looking east, west, north, and south. In Figure 2.2 our observer is positioned at a middle lati-

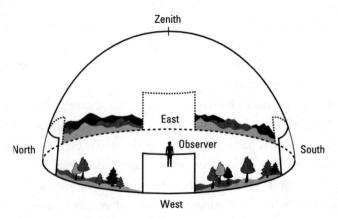

Figure 2.1 The hemisphere of the night sky, showing the four windows of observation used in Figures 2.2 and 2.3.

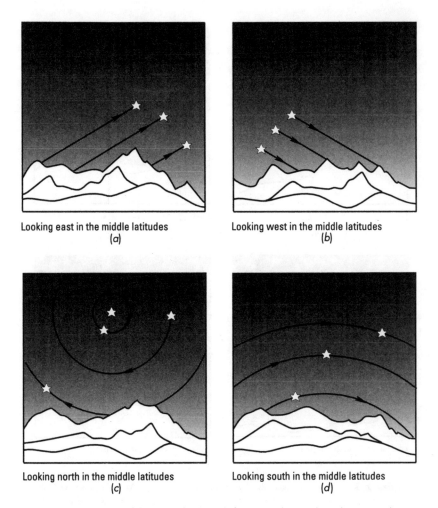

Looking east in the middle latitudes
(a)

Looking west in the middle latitudes
(b)

Looking north in the middle latitudes
(c)

Looking south in the middle latitudes
(d)

Figure 2.2 The view of the night sky as seen from Stonehenge, latitude 51° north, looking in each of the four cardinal directions: (a) east, (b) west, (c) north, (d) south.

tude site, such as that of Stonehenge, while in Figure 2.3, he or she is located somewhere in the tropics, where the view corresponds more closely to what the Maya and Inca people would have seen.

At least three significant differences exist between the sky in tropical (equatorial) and temperate (high latitude) zones that we will need to keep in mind when we examine the Mesoamerican and Andean indigenous systems of astronomy relative to that of Stonehenge.

· First, as you change latitude, sky objects move along paths of different shape and orientation relative to the horizon. In the east and west, daily star

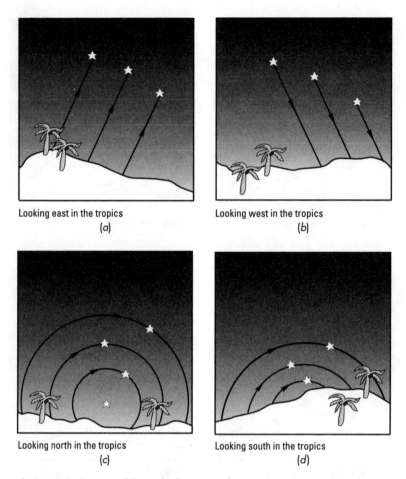

Looking east in the tropics
(a)

Looking west in the tropics
(b)

Looking north in the tropics
(c)

Looking south in the tropics
(d)

Figure 2.3 The view of the night sky as seen from northern Yucatan, latitude 20°
north, in all four directions: (a) east, (b) west, (c) north, (d) south. The Inca sky (latitude 13° south) would resemble the situation in the second set of pictures except
the star trails in parts (a) and (b) would slant in the opposite direction from the vertical, and pictures (c) and (d) would be reversed.

paths are steeper, that is, closer to vertical for an observer near the equator,
and they become more horizontal as your location moves to higher latitudes. For example, compare the angle that the nightly star trails make with
the horizon at our sample latitudes in Figures 2.2 and 2.3. It is 70° or 20°
from the vertical for the Maya, and 77° or 13° from the vertical for the Inca
of Cuzco. If the observer is close to the geographic equator, the angle would
be nearly 90°. For an observer at latitude zero, the stars would plunge ver-

tically into the horizon. Conversely, at Stonehenge the angle of star rise-set is a relatively shallow 39° from the horizon, or 51° from the vertical. In other words, stars rising over Stonehenge move as much to the right as they do upward, setting stars as much to the right as downward. (Over the course of a day or night, the sun, moon, and planets would behave the same way.) By extrapolation, observers in the Arctic and Antarctic regions will see stars rise and set at grazing incidence to the horizon. For an observer at the north geographic pole, nightly star paths would be exactly parallel to the horizon, which means that no star will ever rise or set.

These differences in sky orientation can have major implications for how people view the world. The nearly vertical star paths give the tropical observer a fixed stellar reference direction on the horizon for a longer duration of the night than the temperate zone observer, a fact that figures prominently in the development of Polynesian navigational techniques, for example (see Figure 2.4).

The northern shore of a small Pacific atoll in the Gilbert Islands is dotted with half a dozen pairs of parallel rough-cut slabs, each about the size of a person. They are arranged horizontally and cemented into the ground. One pair points to the neighboring island of Tamana 50 miles distant, another to Beru Island 85 miles away, while a third gives the alignment of distant Banaba, 450 miles over the horizon. Islanders called them "stone-canoes," or "the stones for voyaging." Locals say the slabs were used as instructional models by their ancestors to set directions for interisland navigation. Each pair of stones aligns with the place where certain stars appear or disappear along imaginary lines perpendicular to the sea horizon at different times during the night. For example, in August the bright star Regulus aligns with the Tamana stone at sunset, while at midnight Arcturus gives the same bearing. The navigator simply memorizes a "constellation" that consists of a long vertical chain of stars associated with the selected island, then steers the canoe toward them, holding the mast fixed on one or more stars in the chain. In effect, the star-rise positions become the points of a star memory compass whose elements are transmitted by oral tradition. Such correlations for accurate long-distance sailing must have required many generations of trial and error to develop. But, as Figure 2.4 demonstrates, this technique of navigational star fixing clearly would not work at high latitudes where stars slide right or left as they rise and set.

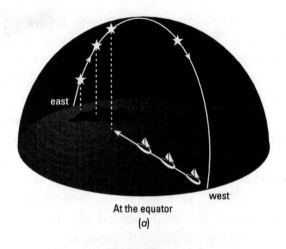

At the equator
(a)

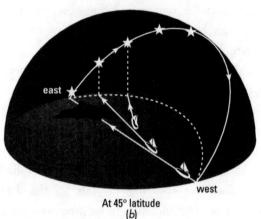

At 45° latitude
(b)

Figure 2.4 The difference in navigation techniques mitigated by nature between equatorial and more northerly or southerly latitudes is shown here. Taking a fix on a star will lead in the desired direction for a sailor close to the equator (a), but in higher latitudes (b), the sailor's course would deviate considerably as the guide star moves laterally across the sky.

Suppose next we look at sky views toward the north and south as opposed to east and west. Stars near the celestial pole—a fixed point on the sky about which the stars move in their 24-hour orbits—appear to sketch out complete circular arcs on the sky every day. Polaris, the Pole Star, today more or less marks the pivotal spot. Because they always lie above the horizon, and therefore remain visible on every night of the year, these stars are said to be *circumpolar*. As a comparison of Figures 2.2c and 2.2d with Figures 2.3c and 2.3d shows, the circumpolar zone of stars occupies a cap of the celestial sphere of greater extent as the observer is stationed at higher latitudes.[1] For tropically based observers, the pivot point lies low in the sky and consequently the circumpolar cap is much smaller. In the southern hemisphere the circumpolar zone is situated in the south, but there is no bright star to mark it as there is in the northern polar zone.

A second difference between tropical and temperate zone skies is that the percentage of celestial phenomena available to the viewer varies with latitude. Basically, there is more to see in the tropics than at higher latitudes. Some stars that are visible in the tropics never can be viewed from the high latitudes. However, in the high latitudes a larger percentage of stars is visible on each night of the year. (Compare Figure 2.2c with Figure 2.3c.)[2]

Finally, another difference between tropical and temperate skies is that the movement of celestial bodies in the skies over the tropics seems to be more closely centered on the observer, while that in the high latitudes appears to be more heaven-centered. This, too, has its cultural implications. In Chinese astronomy, for example, the celestial pole, which is located relatively high in the sky, was regarded as the center of the Purple Palace, the emperor's residence. It corresponded to the emperor himself, the pivot of authority in the state, because it always remained steady and fixed. Closely gathered about him revolved the members of the palace court, each earthly official in the agrarian bureaucracy having been given a named celestial counterpart. By contrast, ancient Aztec astronomers of the tropics designated the fifth cardinal direction as the pivot of the world and the place where time began. It emanated from the center of their waterbound capital city and passed straight up to the overhead point. When the Pleiades passed the zenith of Tenochtitlan at midnight, the emperor sent priests with sacrifices to the Hill of the Star, a prominent mountain that protrudes into the lake basin. Their task was to signal whether a new cycle of time would be granted all mankind. The occasion was celebrated by casting out old housewares and renewing the hearth—in effect, making a New Year's resolution to renew society.

To get the sense of this difference between tropical sky symmetry and high latitude asymmetry, imagine a line connecting the east-west points of the horizon through the observer. Perceived from a tropical vantage point, this line forms a nearly symmetric dividing line about the observer so that all objects are seen to move on the northern side of the hemisphere in the same manner as those on the southern half. This is why the tropical observer seems to lie close to the center of cosmic symmetry. Now look at the four frames of Figure 2.2 for Stonehenge and compare them with the view from the tropics (Figure 2.3). In the former, the pivot of all sky motion is very high in the northern sky, while its corresponding opposite in the southern sky has plunged well below the horizon out of sight; therefore, the eye is automatically directed to that fixed pivot high in the sky to one side of the east-west line.

What you see in the heavens, then, depends on where you are in the world. People who live in the tropics witness the same motion of astronomical bodies in the day and night sky, more or less straight up in the east and straight down in the west. They might be inclined to believe they are at the center of motion because the northern and southern half-hemispheres of the sky behave nearly identically. By contrast, those who live outside the tropics see the center of all sky motion as a fixed point eccentrically positioned high above their heads. Such a remarkable contrast in the changing aspects of the heavens viewed from different places on earth affects everyone's cosmic outlook. Once we begin to single out the bright lights that move in the sky we will discover even more profound differences.

CHARTING THE SUN GOD'S COURSE

A nineteenth-century political chief of the Maya town of Peto, Yucutan, wrote that his ancestors started their year not on January 1 but on July 16, for this was the day on which the sun returned to the overhead point after having spent a brief period of time in northern skies. This point they marked very carefully, as they did the pivotal points on the horizon where the sun reached its annual northerly and southerly limits and the position halfway between them. They accomplished it all by positioning the rising/setting sun between notched sticks as it passed through a natural feature on a neighboring hill. And they did it all to fix the dates of planting and harvesting, says another informant.

While the 24-hour motion of all sky objects is quite regular, over time the sun, moon, and planets behave in a more erratic way. The sun's motion is really twofold. Like all celestial bodies, it always rises somewhere in the east and sets somewhere in the west. But it also appears to move against the background of the stars and constellations of the zodiac from west to east, making a complete circuit of 360° against the stellar background in a tropical year (365.2422 days). *Relative to the stars*, the sun moves about 1°, or twice its own diameter, every day in a direction opposite its daily motion. Two motions that happen at the same time can be difficult to conceive, but here is a familiar analogy that simplifies what happens. Suppose you are seated toward the front of an airplane flying west from New York to Los Angeles. You get up out of your seat and walk at a steady pace down the aisle toward the back of the plane. Your motion back toward New York relative to the seated passengers is like the motion of the sun from west to east relative to the stars. However, the movement of the entire airplane, with you and all the other passengers in it, traced out along the ground by the plane's shadow, can be compared with the 24-hour motion of the sun (along with that of every astronomical body in the sky) from east to west.

One consequence of this double motion is that constellation patterns will appear to shift their positions relative to the sun throughout the year. Because our clocks are geared to sun time, this means that we will see a given star rise a few minutes earlier each day. Thus, if we watch the stars above the eastern horizon opposite the direction of sunset in the darkening twilight on successive nights, new constellations will begin to appear over the skyline. Stars that were visible on the horizon a few weeks earlier will have moved higher up in the sky as the seasons progress, while their counterparts sink out of sight in the west. The times of annual appearances and disappearances, termed heliacal risings/settings (after *helios*, the Greek word for "sun") are among the foremost natural timekeeping devices used by diverse cultures of the world. (These will be discussed in more detail later.) To give an example, shortly after sunset around the time of the winter solstice, our Orion the Hunter, recognizable by the bright stars Betelgeuse and Rigel and the belt and sword in between, appears about a quarter of the way to the zenith above the east point, just as Sirius is about to rise in the southeast. A month later Sirius is well up by dusk and Orion has passed closer to the overhead region by the time it is dark.

A 3,000-year-old recipe for wine making says:

> Then, when Orion and Seirios are come to the middle of the sky, and the rosy-fingered Dawn confronts Arcturus, then, Perses, cut off all your grapes and bring them home with you. Show your grapes to the sun for ten days and for ten nights, cover them with shade for five, and on the sixth day press out the gifts of bountiful Dionysos into jars.[3]

This was Greek poet Hesiod's metaphoric way of telling his brother Perses, who was minding the farm, that once Sirius crosses the local meridian (the north–south line passing overhead) and reaches its highest elevation in morning's brightening twilight (marked by the annual reappearance of Arcturus), this is the time to pick grapes. With corrections for long-term sky movement, the date works out to be the middle of September.

At evening twilight on the spring equinox (March 20), when Orion and Sirius have already passed the local meridian and have begun to descend in the west, Leo, with its bright star Regulus, has just cleared the eastern horizon. In the middle of summer Altair and Vega are the bright stars in the east. By this time, red Betelgeuse and blue Rigel have become lost in the sun's glare. These stars eventually return to view in the early morning sky about mid-July after the sun passes them.

The sun's daily path on selected days of the year at the sites we will encounter is summarized in Figure 2.5. (See Appendix A for an exercise involving a few additional examples.) On the day the sun arrives at the vernal or autumnal equinox, it will rise exactly at the east point and it will set at the west point as shown in all three parts of the figure. Later in the spring, as the sun migrates into the northern celestial hemisphere, it will rise and set closer to the north point. The sun attains its maximum northerly rising and setting points about June 21, the summer solstice (or standstill) in the northern hemisphere (or winter solstice in the southern hemisphere).[4] Having reached its greatest northerly progression, the sun swings along the horizon like the bob of a pendulum and begins to move southward again, crossing the east–west line about September 22. On December

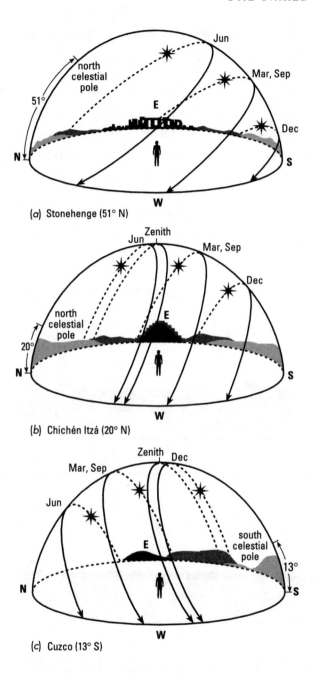

(a) Stonehenge (51° N)

(b) Chichén Itzá (20° N)

(c) Cuzco (13° S)

Figure 2.5 The daily path of the sun on March 20, June 21, September 22, and December 21 as seen from (a) Stonehenge (latitude 51° north), (b) Chichén Itzá in northern Yucatan (latitude 20° north), and (c) Cuzco, Peru (latitude 13° south). Additional lines in (b) and (c) show the path on zenith passage dates.

21 the sun reaches its greatest southerly standstill. At this time it rises and sets nearest the south point of the horizon.

The rhythmic oscillation of the sunrise and sunset points along the local horizon over the course of the seasons afforded many ancient cultures a convenient method of establishing an annual calendar. We can think of the horizon as a calibration device with prominent peaks and valleys that serve as time markers, as shown in Figure 2.6. The sun's rise/set points on the horizon do not change at a uniform rate from day to day. Around the time of the equinoxes, when sunrise and sunset cross the east and west points respectively, the sun's contact points with the horizon shift noticeably from day to day. Near the time of the solstices, when the movement

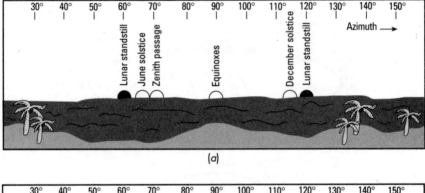

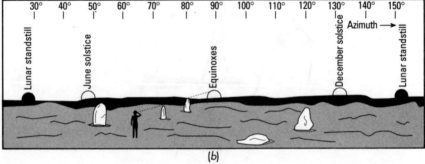

Figure 2.6 The rising sun and moon on pivotal dates as viewed from (a) the tropics and (b) a high northern latitude. Note the increased separation of the standstills at the high latitude site and the nonuniform movement of the sun at the horizon throughout the year at both sites. Darkened disks are placed at the lunar standstills (see p. 33), which are spread wider apart (and slightly exaggerated) along the horizon than the corresponding positions of the sun. Note the rising point of the sun on the days of zenith passage, a phenomenon exclusive to tropical skies. In (b) a hypothetical observer makes a solar alignment with a pair of vertical markers.

of the sun along the ecliptic (the midline of the zodiac) is more nearly parallel to the equator, the change is only very slight. The sun seems to stand still.

Among the Zuni of the southwest United States, a Pekwin (sun priest) was officially designated to follow the course of the sun as it shifted its position on the local horizon. He often built a special sun-watching station at the edge of his village where he would have a clear view. He also kept a calendar on the wall of his house in which he designated important dates by scratches indicating where shadows of the jamb of the sunrise-facing window fell at various times of the year. But the exact date of the solstice sometimes eluded even the most perspicacious Pekwin. In 1881 one sun priest was harshly reprimanded by the council of chiefs for erring by several days in fixing the winter solstice ceremonies. History has a way of smoothing over human frailties and we can only wonder how many times Stonehenge's winter and summer solstice alignments needed to be corrected because some wayward sun watcher failed in his duties.

The cyclic motion of sunrises and sunsets, like that of the pendulum, is perfectly repetitive and, furthermore, it is attuned to the seasons. Next to the day-night cycle and the phases of the moon, the annual oscillation of the sun along the horizon is one of the most universally recognized periods occurring in the natural environment.

One of the most easily recognizable events among astronomies of the tropics is the passage of the sun directly overhead. Solar zenith passages can take place only for observers situated between the Tropics of Cancer and Capricorn.[5] An observer at the Tropic of Cancer would see the sun barely reach the zenith on the June solstice, about June 21. Alta Vista (Chalchihuites), a site in northwest Mexico dating from a time close to the fall of the great Mesoamerican city of Teotihuacan (ca. 600 A.D.), may have been positioned deliberately to mark just such an event and key it in with other points in the solar calendar (see Figure 2.7). Situated well within a degree of the Tropic (latitude 23.5° north), an observer there can still witness the shadowless moment at high noon on June 21 when the sun stands overhead and a stark whiteness overtakes the complex. At dawn on the same day, the sun touches the top of Picacho Peak, a nearby mountain that once may have been regarded as sacred because it marked the source of both precious blue water and the precious bluestones that are mined there. The viewing station,

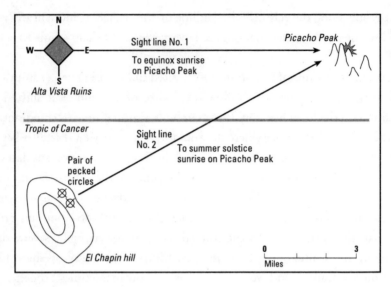

Figure 2.7 A map showing the equinox and summer solstice alignments in the vicinity of the Alta Vista site.

located on a plateau south of the site named El Chapin hill, is marked by a pair of 4-foot quartered circles pecked in stone, which resemble those found at the ruins of Teotihuacan, with their axes aligned to the prominent needlelike peak. Another alignment along a ceremonial walkway within the ruins of Alta Vista marks equinox sunrise over the very same peak. (See Figures 2.8a, 2.8b, and 2.8c.)

North of the Tropic of Cancer, the sun never migrates far enough to the north to reach the overhead point. Likewise, for observers south of the Tropic of Capricorn, the sun cannot move far enough to the south to attain the zenith. For an observer stationed exactly at the Tropic of Capricorn, zenith passage is a single annual event that takes place on the December solstice.[6] At all latitudes between the tropics (23.5° north to 23.5° south), the zenith event happens on two days a year, once when the sun passes the zenith on its northward migration and again when it moves back toward the south. At the geographic equator a pair of zenith passage dates would occur at the equinoxes, nearly equidistant in time from the solstitial dates.

The solar zenith event is well documented in both the Maya and the Inca cultures, and it seems to have played a critical role in the development of their calendars. Spanish chronicler Garcilaso de la Vega is quite explicit about Inca sun-watching methods:

> To ascertain the time of the equinoxes they had splendidly carved stone columns erected in the squares of courtyards before the temples of the Sun. When the priests felt that the equinox was approaching they took careful daily observations of the shadows cast by the columns.[7]

Apparently the astronomers drew lines between these columns and timed the approaching equinox by following the shadows cast by them as they approached the line. At Quito, the northern capital, which happens to lie almost precisely on the equator, the columns would cast no shadow at noon on the equinoxes and so, says Garcilaso, the columns were venerated there to a greater degree because they gave the Sun the seat he liked the best. The people decked the columns with flowers and aromatic herbs and offered the sun god gold, silver, and precious stones. Unfortunately, there is little archaeological evidence to back up what Garcilaso says and no other Spanish chronicler speaks of shadow-casting devices. A good part of his description may have been colored by his Renaissance education and outlook.

If we look back at Figure 2.5, we see that it includes the zenith passage in Maya Chichén Itzá (2.5b) and in Inca Cuzco (2.5c). For Cuzco the dates of zenith passage are February 13 and October 30, give or take a day. Thus, at Cuzco the sun remains north of the zenith at noon for 259 days and south of it for the remaining 106. At Chichén Itzá, a Maya site in north Yucatan, the dates are May 26 and July 20, thus putting the sun north of the zenith at noon for 55 days and south for 310.

The antipodes to these zenith passage dates in the calendar are the dates when the sun passes through the *nadir*, or the point directly underfoot (hereinafter called the *antizenith*). Each solar antizenith date is timed six months apart from a zenith passage date. August 18 and April 26 are the solar antizenith passage dates for Cuzco and January 17 and November 27 are the corresponding dates for Chichén Itzá. The first of these dates was used to establish the starting point of the state calendar of the ancient Inca empire because of a peculiar coincidence. The time to begin planting in the highest altitudes around the capital happened to coincide with the date when the sun

(a)

Picacho Peak

(b)

Figure 2.8 Near Alta Vista (Chalchihuites), a sun watcher's station at the Tropic of Cancer is marked by a carving high atop a flat promontory (El Chapin hill). The axis of the pecked circle (a) points to sunrise on the June solstice over Picacho Peak (b), while the view of the equinox sunrise from Alta Vista's major temple a few miles to the north, enhanced by a ceremonial walkway (c), marks the very same spot on the horizon. Photos by H. Hartung.

set at a point exactly opposite where it rose on the day of zenith passage—the antizenith date in the calendar. Calendar keepers erected a series of pillars to mark the progress of the sun from that point. As it passed each pillar, the sun itself would signal the time to plant at successively lower altitudes. The Inca brilliantly turned the hills immediately overlooking Cuzco into a

Figure 2.8 (Continued).

natural clock for calibrating their agricultural calendar. Later we will explore further the great sensitivity both Maya and Andean people exhibited to these uniquely tropical phenomena of solar zenith and antizenith passage.

THE SUN'S
NIGHTTIME IMITATOR

There was always something odd, something unusual about the ruler of the night sky:

> When he newly appeared, he was like a small bow, like a bent straw lip ornament—a small one. He did not yet shine. Very slowly, he went growing larger, becoming round and disc-shaped.[8]

Once he became large and round like a skillet, the Aztec legend continues, observers could see that the moon has a rabbit stretched across his face (Figure 2.9a). One day when the gods were taunting him, the story goes, they flung a rabbit at him, dimming his face enough so that he could not be confused with the life-giving sun.

Compared with the annual oscillating motion of the sun, the moon's cycle of changes is far more complex. Yet the lunar disk, as the Aztecs long suspected, is the solar opposite, its darker counterpart imitating in mirror image whatever the sun does.

That the second brightest luminary in the heavens passes through phases was surely the most obvious lunar aspect to register in the minds of archaic observers. In Western lore the phase cycle depicts the career of the Man in the Moon (Figure 2.9a). He begins his adventure when the waxing crescent first appears by fighting off the devil of darkness, a dragon, who has eaten up his father, the old moon. No sooner does he reach the height of power at full moon than the old enemy, the devil of darkness who conquered his father, begins to attack and wear him away.

Like the sun and the stars, the moon also rises and sets at the same tilt relative to the local horizon; however, for over half its cycle, the moon is visible for longer periods in the day than in the night sky. Sometimes for a couple of days it is totally absent both day and night. But it is the inconstancy of the way the moon's face looks, the way it changes dramatically and repeatedly, that makes it unique as a celestial body: from a thin sickle to a D-shaped quarter, then to a bulgy egg, and finally, like the golden sun, to a full, if somewhat tarnished, silvery disk. Over the course of a number of days that you could tally on your fingers and toes, you could watch the moon pass through half of its phases from thin crescent to a fully illuminated disk in the sky around dusk. Then you could follow the other half cycle in reverse if you watched the skies just before dawn.

A persistent skywatcher could also chart from memory the course of the moon among the constellations of the zodiac. If you imagine the starry background to be held fixed, you will notice that the moon shifts about a handspan held at arm's length per night from west to east; in other words, it does what the sun does throughout the seasons, only much faster, moving opposite to the normal east-to-west daily motion of the stars across the sky in a month instead of a year.[9]

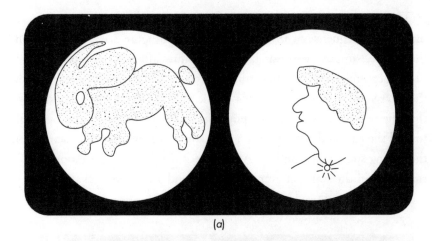

(a)

(b)

Figure 2.9 (a) Figures in the moon: (right) the Man in the Moon of Western lore; (left) the rabbit in the moon, recognized in parts of both Asia and the Americas. The photo of the moon (Figure 2.9b), taken through a low-power telescope, reveals the outlines of the two figures. To see them more clearly, squint your eyes. This helps obliterate some of the distracting detail.

The combined effect of the two lunar motions is illustrated in the month-long sequence of time-lapse drawings in Figure 2.10. To understand this figure, suppose you watch the moon just after sunset from night to night. Beginning with the first thin crescent low in the west hovering above the afterglow of sunset (as shown in the position labeled 1), you discover that the moon takes up a new position, moving that handspan per night toward the east and getting fatter, or waxing, as it widens its spacing from the setting sun (days 2 and 3). After a week the first quarter moon (day 7) rides high in the south at evening twilight and after two weeks, the moon's

(a)

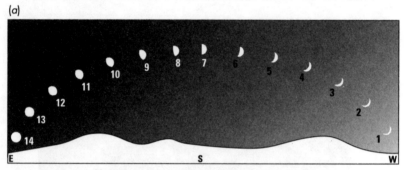

Moon on successive days, looking west after sunset

(b)

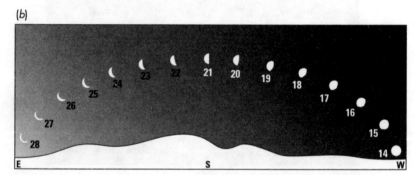

Moon on successive days, looking east before sunrise

Figure 2.10 The many faces of the moon. The moon is the only image in the sky that regularly and visibly changes its shape over the course of its cycle as seen with the naked eye. In about a month the moon moves eastward among the stars all the way around the sky, passing through all its phases as it goes. (a) On days 1 to 14, the observer looks to the south after sunset and sees the moon wax from first visible crescent to full phase. (b) On days 14 to 28, the observer watches the southern sky before sunrise and witnesses the moon wane from full to last crescent. Gone for a day or two, the moon finally returns to position 1.

fully illuminated disk rises in the east, opposite the place where the sun sets in the west. At this time the observer appears to stand on a straight line between the two disks. For the rest of the month, the moon is more accessible to someone who gets up early and views it as its phases wane from full to last quarter (day 21) to crescent (days 26, 27, 28). Finally the thinning crescent vanishes in the sun's glare for a day or two (we call it the "new moon," even though no moon is visible at this time) to complete the cycle.

A passage from a creation myth from the Middle East makes it clear that as early as 1000 B.C., our ancestors were keeping time by every step of this cycle. When Marduk, king of the Babylonian gods, created the moon, he commanded it:

> At the beginning of the month, namely of the rising
>   over the land,
> Thou shalt shine with horns to make known six days;
> On the seventh day with half a tiara
> At the full moon thou shalt stand in opposition to the sun,
>   in the middle of each month.
> When the sun has overtaken thee on the foundation of heaven,
> Decrease the tiara of full light and form it backward.
> At the period of invisibility draw near to the
>   way of the sun,
> And on the twenty-ninth thou shalt stand
>   in opposition to the sun a second time.[10]

Though the phase cycle seems to have been recognized universally, ancient astronomers recorded it in different ways. Usually, it was tallied as a 29- or 30-day period beginning with the first appearance of the thin waxing crescent. Ancient Greek astronomers stood on the roofs of buildings at sunset to spot the thin-lined crescent that would begin the count. (Imagine the monthly payments on your credit card depending on a time cycle initiated by such a sighting!) In other cultures the month was counted from new to new and even from full to full phases. Modern astronomers measure the moon's phase cycle to be 29.530589 days long and they call it the *lunar synodic month*. The Maya succeeded in charting the lunar synodic month with great accuracy. Among the American Indians of the U.S. Southwest, only the days on which the moon was *visible* were named; thus the number 28, not 29 or 30, appears as their most important lunar count.

We call the first full moon following the autumn equinox the harvest moon, and hunter's moon the one after that. Both are vestiges of a lunar-cycle naming system that has survived the Native Americans who lived in the continental United States before the time of European contact. Many of these tribes had a name for each of the twelve (sometimes thirteen) months of the year. The Natchez tribes, hunter-gatherers of the Mississippi River Valley for example, had a Deer month, Strawberry month, Little-corn month, Watermelon month, Peach month, Mulberry month—even a Nut month (when they crushed nuts and mixed them with flour to make bread). Tribes in the northwestern United States and Canada, on the other hand, celebrated a Salmon-Spawning month, Pine-Sapwood month, and a Ducks-and Geese-Go-Away month. By reading these lists of names we learn a lot about the people who kept time by the phases of the moon. We can almost sense the flow of seasonal rhythm among people for whom time is activity and experience itself, rather than a mechanical device designated by the ticks of a clock, as we in the industrialized world have come to view it.

There is another kind of month—one measured by the stars—which will turn out to be of special importance when we discuss the Inca calendar. Remember the star-studded corridor that circles the sky—the one the sun follows in a year and the moon treads in about a month—the *zodiac?* The word means "circle of animals." The constellations that mark out this band, many of them depicted by the figures of animals, have received much attention in legend and folklore because they chart the courses of not just the moon and the sun but also of five other bright celestial lights: the planets Mercury, Venus, Mars, Jupiter, and Saturn. The time it takes the moon to return to the same place among the stars is called a *lunar sidereal month,* from the Latin word *sidus,* meaning "star." Modern astronomers have determined the length of the sidereal month to be 27.32166 days, or a little over two days short of the synodic month. Thus, if the full moon appears in a certain position in the constellation of Taurus this evening, you can be sure that it will return to the same position after approximately 27⅓ days (though it will not quite yet be full). However, because of the fractional number of days involved in the sidereal period, the moon will resume that position one-third of a day (eight hours) later in the evening. If you make this evening's observation at midnight, you should make the next observation one sidereal month later at eight a.m.; however, at this time the sun already will have risen and the background stars will no longer be visible. From a practical point of view, it might be convenient to think of sidereal months as occurring in groups of three. Thus, after a period of 3 × 27⅓ = 82 days (81.96498, to be precise), the moon will assume its original position in Taurus *at the same time of the night.* When charting the motion of the moon among the constellations, there is reason to think that some ancient astronomers may have assigned greater importance to the 82-day cycle as a matter of simple convenience than to the 27⅓-day interval. In fact, Quiché Maya diviners of the highland Guatemala town of Momostenango call this period the *chaé alic,* meaning "staked" or "stabilized," and they use the opening day of each cycle to predict rain. They make visits to hilltop shrines overlooking their town in each of the four cardinal directions in accordance with a pair of cycles that coordinates four overlapping 82-day periods with the four divisions their 260-day sacred almanac.[11]

While the moon follows the general course of the sun through the zodiac, it deviates from the ecliptic, the midline of this celestial band, by a little more than 5° (5°09', to be exact). This means that there are occasions when the moon can migrate 5° farther north and south compared with the sun; and so, the moon can rise and set farther north and south of the sun's standstill positions or solstices. Moreover, these *lunistices* stretch farther north and south from the east-west line as the observer frequents them at higher latitudes, where the risings and settings occur ever more obliquely to the horizon. At Stonehenge the standstills are more than 90° apart; Stonehengers may have made use of these principles in the design of their ceremonial center. Figure 2.6 illustrates this effect.

Suppose that the sun is at the June solstice when the moon attains its northern standstill. Then the southern limit, which would coincide with a new or full moon at the December solstice, would place the moon at its maximum southerly standstill.[12] The moon reaches a given standstill every 18.61 years, another fact that seems to have been known by the builders of Stonehenge. (Extremes last occurred in late 1987 and will not b reached again until sometime in early 2006.)

With some understanding of the difference between the synodic and sidereal periods of the moon, we are able to analyze the workings of one of the grandest of nature's phenomena: eclipses. Of all the spectacles nature puts on for us, none is so breathtaking, so out of the ordinary, as the darkness in the daytime that accompanies an eclipse of the sun. (The quickening pace of the phase cycle—all in a night—accompanied by an eerie, coppery, or blood-red eclipsed full moon is a close second, in my opinion.) Little wonder eclipses were often regarded as omens of disaster. As early as 2000 B.C., a pair of eclipses of the moon were said to be connected with the murder of the Mesopotamian king of Ur and with the destruction of his city. And the chronicler of the Aztecs, Father Bernardino de Sahagun, tells us what happened in Mexico on the occasion of that rare daytime darkness.

> When the sun is eclipsed—they then raise a tumult—and then the women weep aloud. And the men cry out [at the same time] striking their mouths with [the palms of] their hands. And everywhere great shouts and cries and howls were raised. And then they hunted out men of fair hair and white faces; and they sacrificed them to the sun. And also they sacrificed captives, and they anointed themselves with the blood of their [own] ears. And, besides, they bored [the lobe of] their ears with maguey spines and passed flexible twigs or the like through the holes which the spines had made. And then, in all the temples, they sang and sounded [musical instruments], making a great din. And they said: "If the sun becometh completely eclipsed, nevermore will he give light; eternal darkness will fall and the demons will come down. They will come to eat us!"[13]

If the moon's course about the earth coincided with that of the sun's path along the zodiac, an eclipse of the moon would occur at every full moon, when the observer aligned exactly with the sun and moon disks. (At this time the moon would fall within the shadow of the earth.) Solar eclipses would occur two weeks later at new moon when the moon lay between the observer and the sun. However, as we have already indicated, the course of the moon is inclined by 5° to the ecliptic. This changes the picture considerably.

Eclipses occur only when a new or full moon lies close to one of the two intersection points (astronomers call them *nodes*) of the sun's and moon's paths across the zodiac. This happens periodically, but because the moon and the sun (or the moon and the earth's shadow, if we are dealing with a lunar eclipse) are disks rather than points, they can cross slightly to one side or the other of a given node and still produce a partial eclipse. The situation is depicted in Figure 2.11. That eclipses repeat in cycles extending over long periods of time was one of the most absorbing discoveries made by our predecessors. Finding complex patterns that emerge from the data of observation must have required persistent sightings and individual genius.

How to predict eclipses? Basically, when whole multiples of lunar and solar periods fit together, eclipses must repeat. Suppose, for example, that a full moon at one of the nodes gets eclipsed. One synodic month (29.53059 days) later, the moon is full again, but by that time the moon already will have passed the node. The time it takes the moon to return to a given node is 27.21222 days.[14] It is named the *draconic* month, after the dragon the ancient Chinese believed devoured the sun or moon during an eclipse. Eclipses recur only when these two cycles are synchronized. For example, if these periods were simple whole numbers, say 30 and 27 days respectively, then an eclipse would take place every 270 days. This is because nine full moons (9 × 30) would equal exactly ten passages of the moon by the node (10 × 27). (Actually, a half-integral multiple of the draconic month will do, since the moon may be eclipsed at the opposite node.) In reality, there is no number fulfilling the desired conditions for generating eclipse cycles per-

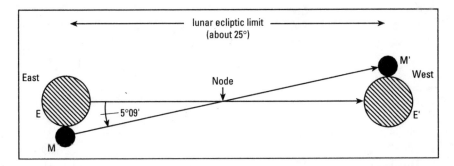

Figure 2.11 The region of one of the nodes of the moon's orbit (MM') magnified to show the zone along the ecliptic (EE'), or the sun's (and the earth shadow's) path along the zodiac, within which an eclipse of the moon can take place. The lunar ecliptic limit extends twelve or thirteen degrees either side of node passage. It takes the shadow about 25 days, the moon about two, to traverse this zone. (The shaded area represents earth's shadow; the black disk is the moon.)

fectly, but some long intervals do come close. Table 2.1 lists several of them. They were derived simply by searching for close matches of whole multiples of synodic months with whole and half-whole numbers of draconic months.

Any of these intervals could have been recognized by the ancients, though some cycles might have been more easily detectable than others. Some astronomers have argued that eclipse-prediction cycles were incorporated into the architecture of Stonehenge. And as we shall see, several eclipse cycles may have been known to the Maya. The most famous of all the eclipse periods in Western lore is the 6,585.32-day cycle. It was discovered by the ancient Assyrians about the fourth century B.C. and later named the

TABLE 2.1. A FEW COMMON ECLIPSE CYCLES

Days	Cycle Years	Number of Synodic Months	Number of Draconic Months
1,033.57	2.83	35	38.0
1,210.75	3.31	41	44.5
1,387.93	3.80	47	51.0
1,565.12	4.29	53	57.5
1,742.30	4.77	59	64.0
2,067.14	5.66	70	76.0
2,244.32	6.14	76	82.5
2,421.50	6.63	82	89.0
2,598.69	7.11	88	95.5
2,775.88	7.60	94	102.0
2,953.06	8.09	100	108.5
3,130.24	8.57	106	115.0
3,986.63	10.92	135	146.5
6,585.32	18.03	223	242.0
9,184.01	25.14	311	337.5
10,571.95	28.95	358	388.5
11,959.89	32.74	405	439.5
14,558.58	39.86	493	535.0
17,157.27	46.98	581	630.5
18,545.21	50.78	628	681.5
19,755.96	54.09	669	726.0
22,531.84	61.69	763	828.0
23,742.59	65.01	804	872.5
25,130.53	68.81	851	923.5
27,729.22	75.92	939	1,091.0

saros, which means repetition. As Table 2.1 indicates, counting forward 223 new moons from the time of a solar eclipse (column 3), we would expect another solar eclipse to occur, because the *saros* interval is also a whole number of draconic months, 242 to be exact (column 4). If the first eclipse took place exactly at the node, the second would occur shortly before the moon arrived at the node. This is because 242 draconic months is actually about one hour longer than 223 synodic months; therefore the new moon will be one hour (about ½°) short of arrival at the node on the occasion of the second eclipse. The third eclipse in the series, 223 synodic months later, would occur a full degree (two lunar diameters) west of the node. After about thirty-five eclipses, the moon would slide off the end of the region about the node within which eclipses are possible, thus terminating a particular series. The *saros* probably attracted further attention because it is also nearly equal to a whole number of years (column 2)—again two cycles that fit together perfectly. This means eclipses in the *saros* series are seasonal— that is, they occur at approximately the same time of year, backsliding an average of 11 days per cycle. Recognizing such long-term patterns of eclipses like those indicated in Table 2.1 would have called for far more attentive skywatching than that required to detect patterns of solar and lunar standstills. Consequently, many scholars argue that cultures that utilized these periods in their astronomical calendars also must have developed sophisticated notational systems—written records to pass such knowledge on to their descendants. This, too, will turn out to be a bone of contention in our study of ancient astronomies.

RETURNING WANDERERS

The Greeks called the planets "wanderers" after the Babylonians who characterized them as sheep who escape from the fold—presumably the rest of the stars. This is because they move against the natural grain of the east–west daily motion of the stars. Red Mars, bright white Venus, swift Mercury, and slower-moving Jupiter and Saturn—each possesses its own unique track across the zodiac. And each was a reigning deity endowed with human characteristics to suit its own peculiar behavior. In Old World astronomy, phlegmatic Saturn was an old man because he moved slowest of all around on his orb. Venus (Ishtar to the Babylonians) represented a fickle *femme fatale* because she constantly courted the sun god, tantalizing him by brushing by, even venturing into the underworld in pursuit of him, only to flee to a place high in the sky after each affair. An old hymn lauds her beauty:

> *Ishtar is clothed with pleasure and love*
> *She is laden with vitality, charm and voluptuousness.*
> *In lips she is sweet; life is in her mouth.*
> *At her appearance rejoicing becomes full.*
> *She is glorious; veils are thrown over her head.*
> *Her figure is beautiful; her eyes are brilliant.*[15]

In the middle of the hierarchy of motion lay bright Jupiter, who moved neither too fast nor too slow. He was moderate, fair in judgment—an appropriate king of the gods.

In addition to the west-to-east motion among the stars, which they share with both the sun and the moon, each planet also undergoes retrograde motion. On its course through the sky relative to the stars, a planet appears to slow down, come to a stop, and for a period of several days to a few months to move backward, that is from east to west. Then it slows to a stationary position a second time and resumes its normal west-to-east motion again. Retrograde loops are particularly prominent for Mars, Jupiter, and Saturn; they are tight and fairly small. Venus and Mercury, by contrast, spend almost as much time going backward as forward among the stars.

Figure 2.12 shows what you would see if you followed the motion of Mars against the stars for several weeks. Recognition of the retrograde loop shown in the figure has had an indelible impact, at least on Western astronomy, from the time of the Babylonians and Greeks all the way up to Copernicus and the Renaissance, but it also may have made an impression upon the Maya. As we shall see, the ancient Greeks and their descendants dreamed up various geometrical models to explain why planets appear to wander among the stars and especially why they make retrograde loops. The seventeenth-century German astronomer Johannes Kepler spent a significant portion of his life trying to mathematically work out the orbit of Mars

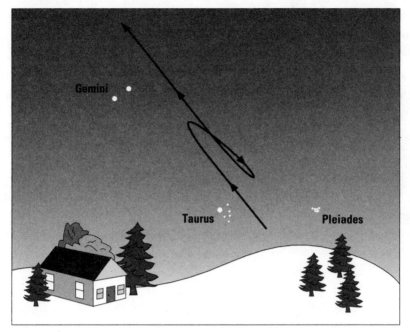

West horizon

Figure 2.12 Here the retrograde motion of Mars is shown in a time-lapse sequence relative to the background of stars over several weeks. First, Mars glides eastward along the zodiac like the sun and moon. But then, just as it is about to enter Gemini from Taurus, it halts, turns back to the west for a few weeks, rests again, then resumes its eastward movement to complete its loop among the stars.

to account for its ever-changing retrograde path. He was finally forced to conclude that Mars must travel on an ellipse and not a circle—a difficult ideological pill to swallow, for the circle had been regarded as divine since the time of the ancient Greeks.

If you want to follow the planets through their complete cycles, that is from one retrograde loop to the next, a month, even a year out under the stars usually will not do. But if you are patient enough to allow yourself (and your descendants!) years of collective experience that can be passed on, it might eventually dawn on you that the five wanderers seem to be of two very distinct types. On the one hand, free-ranging Mars passes all the way around the nighttime sky in about two years. Jupiter and Saturn do the same in a little more than a year. Sometimes they are seen as morning stars, close to the sun in the eastern predawn sky, and at other times as evening stars, setting in the west after the sun. Often they are situated high in the sky at midnight. At other times they disappear from view, lost in the glare of the sun.

In stark contrast to Mars, Jupiter, and Saturn, Mercury and Venus always stay close to the sun; Mercury never strays more than two handspans, Venus a little more than three from the sun's golden light. Over time, each seems to bob like a yo-yo, back and forth, toward and away from the sun, Mercury completing a cycle about three times a year, Venus in a little over a year and a half. The effect is especially noticeable for Venus, first because it is so bright (it is the brightest object in the sky after the sun and the moon) and second because Mercury, Venus's imitator, spends much more time lost in the glare of the sun. In Babylon, the love goddess's aspect, as we might expect, often was accompanied by omens of abundance, fertility, and sexual fulfillment:

> *When Venus stands high, pleasure of copulation.*
> *When Venus stands in her place, upraising*
> *of the hostile forces, fullness of the women*
> *shall there be in the land.*[16]

Venus was special to the Maya, too; therefore it would benefit us to track a full cycle of its motion, the details of which their astronomers were fully aware.

Suppose you are standing under a clear sky looking east in the midst of morning twilight. As time passes and the sky grows brighter, stars lift out of the horizon and glide sideways off to the south. Just as the last star fades from view in the reddened sky, you glimpse a bright light clearing the horizon near dawn's glow. Then in a moment it is gone, swallowed up by the advancing daylight. This is Venus's first appearance as morning star—its heliacal rise. Next morning you return at the same hour. This time Venus rises a few minutes earlier and moves a bit higher in the sky before it vanishes and the sun takes its place. Next day it is still brighter, still higher, and it is visible still longer.

A typical track showing how Venus moves in the morning twilight over several months is shown in Figure 2.13. Round symbols depict where it was last sighted in the vanishing morning twilight at approximately two-week intervals. Dashed lines connect it to the corresponding sunrise position.

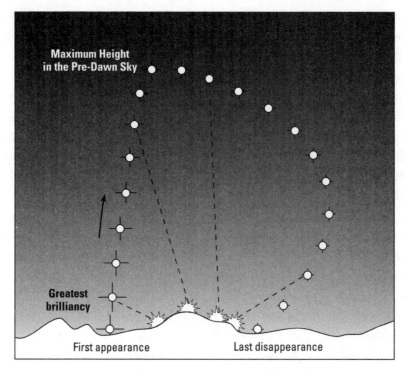

Maximum Height
in the Pre-Dawn Sky

Greatest
brilliancy

First appearance   Last disappearance

Figure 2.13 The quintessential paths of Venus. Standing in the same spot and following the motion of Venus from night to night at the same time of night, this is what you would see. Symbols spaced at two-week intervals show where morning star Venus would just disappear at the end of twilight in the east over several months. As Venus makes a sweeping arc across the twilight sky from first appearance at horizon, through maximum altitude, to last disappearance below the horizon, the sun, connected to it by a dashed line, creeps slowly along the horizon toward its December standstill. The sizes of the spikes on the Venus symbol denote relative brightness. This looped track is one of five distinct shapes the course of Venus can take in the morning or evening sky.

Notice how every day during this period, morning star Venus seems to faithfully announce the sun's arrival; it begins its morning aspect by dramatically bursting upon the celestial scene—that initial appearance out of the solar glare we referred to above. Spokes represent relative brightness, the maximum occurring several days after heliacal rise. From morning to morning Venus gradually stretches away from the sun until it reaches its maximum extension, when it stands at its highest point in the predawn sky. Then it seems to snap back toward the sun and is visible in morning twilight ever more briefly; finally it makes its last morning appearance, disappearing once again in the solar glare. Gone from the predawn sky for several weeks—an average of 50 days—it repeats the process in the evening sky, hovering like a guard dog in the encroaching darkness near the spot where the sun went down. Then Venus acts as an evening star, going from first appearance to last in an average of 263 days, the same as it does when it is the morning star. This time the great white light is out of sight for an average of eight days before it returns to the morning sky and repeats the whole process 584 (263 + 50 + 263 + 8) days later. We call this interval the *synodic period of Venus*. (The planets' synodic periods and their subdivisions are listed in Table 2.2.)

This 584-day synodic cycle, also called the Venus year, just happens to mesh perfectly with the length of the year of our seasons, or 365 days, in the ratio of 5 to 8. This means that to the careful eye any visible aspect of Venus timed relative to the position of the sun will be repeated almost exactly after eight years. For example, if Venus first appears as morning star on the first day of summer 1997, it will do the same very close to that date in 2005.

A seasonal index like this one (or the *saros*) could be useful to any practical-minded people who kept time by a solar-based calendar, especially if they were attracted to expressing periods in whole-number ratios. As we shall discover, the ancient Maya of Yucatan were just such a culture. For a time they became obsessed by the Venus cycle, recognizing that this curious four-phase motion (appearance, long disappearance, appearance, short disap-

TABLE 2.2. OBSERVABLE PERIODS (IN DAYS) FOR THE NAKED-EYE PLANETS

	Synodic Period	Mean Disappearance Intervals	Mean Intervals as Evening/ Morning Star	Mean Interval in Retrograde
Mercury	115.9	5, 35	38,* 38*	—
Venus	583.9	8, 50	263, 263	—
Mars	780.0	120	660	75*
Jupiter	398.9	32	367	120
Saturn	378.1	25	353	140

*These intervals fluctuate widely but commonly lie within about ten days of the quoted values.

pearance) harmonized with a host of other natural periodicities, including the time between the conception and the birth of a child. The Venus eight-year cycle also equals a whole number of lunar synodic months (99 of them to be exact). This means, for example, that whatever phase of the moon accompanies the first appearance of Venus, say at the June solstice in 1997, that same phase will repeat again at about the June solstice eight years later, in the year 2005, signaling the return of Venus.

The coming together of cyclic periods may seem unimportant to us; for example, it scarcely matters what day of the week coincides with New Year's Day from year to year; but as we shall see, for societies whose systems of timekeeping were based on repetitive natural cycles, some of them going all the way back to their mythic creation, the revelation of such fitting together, or "commensurations," in the wandering of their celestial deities would have constituted major discoveries capable of disclosing the secrets of the gods. For example, the Indian calendar tallies a 2,850-year creation cycle made up of 150 Metonic cycles (the time it takes to return a given phase of the moon to the same date of the year), and an even longer cycle of creation, said to have occurred at midnight on February 17–18, 3100 B.C., based on combinations of synodic cycles of the planets. The Maya, surprisingly, put the zero point of their calendar only a decade earlier (August 11, 3113 B.C.), though scholars have yet to work out precisely which celestial (or other?) cyclic beginnings came together on that date. Finally, the Julian calendar by which modern astronomers still operate is based on the unlikely commensuration of three cycles: the length of time over which all possible combinations of days of the week can coincide with the first of the year (28 years), the 19-year Metonic cycle, and a 15-year cycle relating to the imposition of taxes paid to troops upon being discharged from the military of the Eastern Holy Roman Empire—total 7,980 years!

Venus reveals further secrets when we follow it over a full eight-year cycle of morning and evening star curves like the one in Figure 2.13. A set of consecutive curves, such as those shown in Figure 2.14, shows the change of position of Venus as it is observed at last appearance in morning twilight or first appearance in evening twilight over the course of several months. Imagine you were viewing Venus in a series of these time-lapse sequences from a plaza fronting a sacred building in an ancient Maya ceremonial center. If you watched Venus over several years, you might come to realize that the serpen-

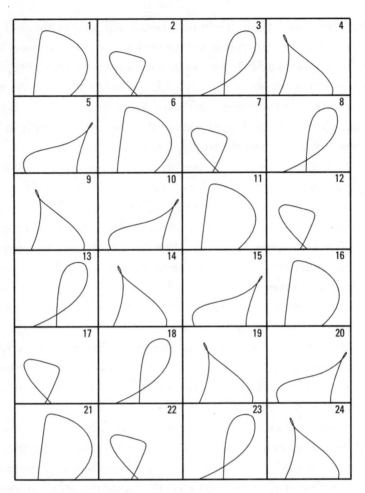

Figure 2.14 A sequence of morning star Venus curves like the one shown in Figure 2.13. Can you recognize the pattern?

tine curve shown in the figure can take on various looped shapes. And, if you compared them, you would find that there were only five repeatable forms of the course of Venus. Can you find the one shown in Figure 2.13 in the full set given in Figure 2.14? The interval between the execution of a pair of sky paths of similar form seen at the same time of night is eight years or 2,920 days, and it is made up of five Venus cycles, each averaging 584 days. These Venusian visual perceptions, especially the grouping of Venus cycles in fives, largely obscured in contemporary astronomy, were indeed noticed by many of our Old and New World predecessors.

Another directly observable planetary cycle, also unrecognized in modern astronomy, has to do with the changing position of the planets as they appear and disappear over the horizon from day to day. As we might expect, Venus executes an annual horizontal oscillating motion that more or less follows the sun, and the standstill limits that it reaches along the horizon. Venus standstills, like those of the moon, are spaced a bit further apart than the solstices and they repeat every eight years. Likewise, the other planets also oscillate like pendulums, each with its own amplitude and period. But because Venus is always near the sun, we might well suspect it was the one watched most closely. This back-and-forth motion, as we shall learn, goes a long way toward explaining references in ancient inscriptions to seasonal phenomena ruled by Venus. In Mayaland, for example, a male Venus deity is tied to rain prediction and the growth of maize. But how could a farmer possibly associate the behavior of Venus and the welfare of his crops? Rainy seasons occur on a yearly basis, but the 584-day Venus cycle is more than a year long. We have a choice: Either the connections they made were purely imaginary, or they saw Venus behave in a way that has escaped our attention.

The fault lies in ourselves and not in the stars! Perspicacious observers who regularly get a good look at the local skyline will discover, perhaps to their surprise, that the *place* of disappearance (or appearance) of Venus at the horizon can easily be correlated with the *length* of its shorter disappearance period, and this in turn is linked to the season of the year. Thus, in the Maya tropics when Venus disappeared during the month of August, it could remain absent from view for more than twenty days on the average, whereas in February it was likely to be gone for only a few days until its morning return. In rare circumstances it is even possible to view Venus as evening and morning star on the same day. But if we average out disappearance intervals of Venus over several years, the result is about eight days, precisely the period assigned to that particular aspect of the planet in the ancient Maya calendar. So we should not be surprised to discover in ancient remains connections and correlations not recorded in high-tech astronomy books of the contemporary age. It is all a matter of what people felt they needed to pay attention to.

As we can see, naked-eye astronomy is all about trying to foretell the future. Our forebears strove to predict well in advance where powerful celestial forces, the harbingers of things to come, would be positioned. Only then could the people prepare themselves. For example, the cycle of Venusian horizon appearances could have provided clever sky watchers with a precise device for predicting things to come that had implications reaching well beyond setting the time for planting or anticipating rain. If the time of appearance and position of Venus at the horizon vary in a regular way with the season of the year, those who bear such knowledge have the capacity to predict exactly how long it would be before Venus would return once it had vanished, as well as where it would be seen when it did return. Knowledge is power, and the corpus of ancient inscriptions, iconography, and architecture will stand as testimony to an abiding concern for detailed sky watching.

STARRY ROADWAYS

Among contemporary farmers in the town of Misminay, Peru, not far from the old capital city of Cuzco, observations of the Pleiades, called by the natives the *collca* or storehouse, are still used to set up a complex agricultural calendar. Anthropologist Gary Urton, who lived among them, learned that farmers make observations of the morning first appearance and evening last appearance of this conspicuous cluster of stars, and that on both occasions they make crop predictions.[17] The farmers say that when the Pleiades reappear large and bright the crops will flourish, while a small and dim rising portends a bad yield. Urton also discovered that planting and the harvest are closely timed by the two observations. As we shall see when we discuss ancient Andean astronomy, the interval in between these observations, which brackets the approximately month-long period during which the Pleiades are absent from the heavens, was the key to the ancient Inca calendar. The Inca started the agricultural year cycle with the appearance of the Pleiades and ended it with their disappearance, not even bothering to count the time when the star group was out of view.

We already learned that because the sun moves among the stars there will be extensive periods during the year when stars and constellations are blocked from view by the solar glare. But how can the dates of disappearance and return of the stars be used as timing mechanisms to fix important civil, religious, and agricultural dates in the year? We know, too, that people also paid attention to the appearance and disappearance of stars on horizons opposite the sun. The so-called heliacal rising and setting events are four in number:

1. the first day on which a star is visible rising in the east before sunrise,
2. the last day on which a star is visible setting in the west after sunset,
3. the first day on which a star is visible rising in the east after sunset, and
4. the last day on which a star is visible setting in the west before sunrise.[18]

To understand how these phenomena work, suppose we track the sun's cycle among the stars in the vicinity of the Pleiades, for an observer based in the Inca capital of Cuzco (latitude 13° south) in the fifteenth century. Observations show that the brightness of a star relative to the skyglow in the vicinity of the horizon along which it rises will determine when that star will first be observed. It was precisely this date to which Greek poet Hesiod was referring when he timed the grape harvest by watching the rosy dawn confront Arcturus. Under the best conditions—that is, in the absence of morning haze—a dense clustering of moderately bright stars like the Pleiades would become visible when the sun is 16° or 17° below the horizon. Thus, the heliacal rise (event 1.) date for the Pleiades in Cuzco would be found to occur about June 4. After this date the Pleiades move out of the range of the region of the glow of dawn and the star group becomes more prominent in the morning sky, rising progressively earlier than the sun each day. About half a year later, the Pleiades will have moved halfway around the sky relative to the rising sun. They take up a position near the *western* horizon as dawn twilight begins to occur in the east. At this time *heliacal setting* (event 2.) occurs, about November 7.

Now suppose that we witness a sunset on this date. Which stars will appear opposite the sun in the east? Because the Pleiades were visible low in the west at the beginning of morning twilight, they cannot yet be seen on the eastern horizon at the end of evening twilight, though the constellations we call Triangulum, Aries, and Perseus, all located slightly west of the Pleiades, should appear low in the east. A few weeks later (about November 18) the Pleiades will undergo another heliacal rising, this time in the evening when they first become visible in the eastern sky after the sun has set in the west (event 3.). During late fall and early winter evenings, the cluster of stars advances ever higher in the eastern sky. Finally, the sun begins to close in on the Pleiades, and about April 18 it will have moved back to within 16° to 17° of the group. Then the Pleiades undergo another heliacal setting (event 4.), this time being barely detectable above the western horizon at the end of morning twilight. Now, though they set after the sun, the little compact cluster becomes invisible because of the solar glare. The Pleiades are lost from view for a period of about a month, approximately May 3 to June 4, when they make their celestial appointment, undergoing heliacal rise, and another Pleiades cycle is resumed.

All over the world astronomers recognized star patterns in the sky. Orion, the brave hunter, done in by the sting of a scorpion, was awarded a prominent place in the late autumn sky where he still goes hunting among the stars—albeit on the opposite side of the firmament from Scorpio, his nemesis. The Big Dipper is the hind end of a great bear, Ursa Major, a beautiful woman transformed and placed by Jove at the insistence of his jealous wife, Juno, in a never-ending circuit close to the celestial pole where Juno could keep an eye on her. There is no reason to think our constellations of Orion and the Big Dipper, obvious patterns to us, would necessarily be recognized by other cultures. Coincidentally, in Iroquois star lore the Big Dipper is also a bear. Pierced by the hunter's arrow at the time when he skims the north horizon, the blood he sheds is said to make the autumn leaves turn red. (An exercise in star pattern recognition is given in Appendix A.)

Nor should we expect that the zodiac of Western astrological fame should be universal. Our modern zodiac consists of twelve constellations or star groups whose origins can be traced all the way back to Egypt and the Middle East. The Sumerians called it "the way of Anu" (the sky god) because it was lit by bright stars that were the fixed counselor gods who surveyed the goings-on in their respective domains and advised the planetary wanderer gods and the sun and moon gods who passed by concerning future undertakings on their journey. Some of the constellations depicted on the ceiling of an Egyptian king's tomb in Figure 2.15a are still recognizable by modern sky enthusiasts. Like the celestial *collca* of the Inca, most of these star groups probably once had meaning as both timing devices and omen-bearing events in the annual calendars. For example, it is no accident that the three watery signs (Capricorn, Aquarius, and Pisces of our zodiac) follow one another. This is probably because the sun once was stationed in those constellations during the rainy season. Also, a Hebraic month such as Sacrifice (*Nisan*) reflects traces of activities that once took place at specific stations in the yearly round. The Maya zodiac, a section of which is depicted in Figure 2.15b, consists of thirteen constellations (12 and 13 are the closest whole multiples of the number of lunar synodic periods in a solar year). Though there is considerable disagreement over the identity of these star groups, it should come as no surprise that the tropical animal zoo is a bit different from that of the ancient Middle East (though surprisingly, the scorpions may be the same!).

There is a second roadway of the stars, often neglected by Western culture but particularly important among people who live in the American tropics, especially because it is so prominently visible from the latitudes in which these cultures developed. The Milky Way is visible as a 10°-15°–wide band of diffuse light that passes all the way around the sky at about a 60° angle to the daily direction of motion of the stars. It is best viewed when it crosses the zenith from north to south on late summer evenings. The Maya thought of it as the umbilical cord that connected heaven and the underworld to the earth and as the cosmic tree of life. Some contemporary Maya groups still regard it as a great celestial roadway. The Chorti Maya, for example, call the Milky Way the *Camino de Santiago* (the Way of St. James) after the old Spanish usage. They pay particular attention to its situation in the sky relative to the position of the sun. The Milky Way is also described as a celestial river that carries the paddler gods and First Father, our creator, into and out of the underworld, *Xibalba*. Maya epigrapher Linda Schele has discerned a pattern in Maya iconography that can be related directly to the *Popol Vuh*, or the story of creation as told in the sky, in which the Milky Way provides the background set for the tale.[19] In this scenario the ecliptic is a two-headed serpent and the constellations positioned at the location where it crosses the Milky Way are especially important. The imaginative story begins with the lighting of the hearth (the lower region of Orion including the Orion nebula, see Figure 2.15b) by First Father, who is reborn out of the shell of a cosmic tortoise (the belt of Orion). He raises the great World Tree

(a)

(b)

Figure 2.15 Star patterns (constellations) recognized by various cultures of the world.
(a) Egypt, Constellations of the northern sky depicted on the ceiling of the tomb of fourteenth-century B.C. Egyptian King Seti I. (b) Maya sky peccaries (our Gemini), tortoise, and triangular hearthstone (lower Orion) mark the region where the ecliptic and the Milky Way, the two great celestial roads, intersect.

(the Milky Way), which is in the shape of a crocodile. The Pleiades, which come next, represent the handful of seeds which, when planted, grow into the fertile World Tree seen passing prominently through the north-south-overhead zone a few hours after the Pleiades have crossed the zenith.

What is so compelling about this Genesis-in-the-sky story is that so many of these sky patterns and events are bolstered by what we find in ancient Maya writing and sculpture. While scholars are not in complete agreement about details of the sky story, this much is clear: For the Maya

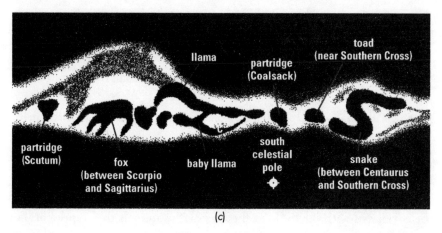

Figure 2.15 (*Continued*). (c) Contemporary Andean dark cloud constellations of the southern Milky Way; to the people of Misminay, dark cloud constellations represent animals.

horizon astronomy was only a part of the picture. What happened high up in the sidereal heavens mattered, too.

Civilizations of the Andes also employed the Milky Way for basic orientation as well as for timing events. They also harbored the idea that celestial objects represent a continuation of things terrestrial; thus, the Milky Way is *Mayu*, a celestial roadway. Pilgrimages tied to the location of the Milky Way along the northwest-southeast (intercardinal) axis at a particular time of the year, such as the June solstice, have been traced all the way back to precontact times when Inca priests followed the route from Cuzco to the sacred site of Vilcanota along that direction; they literally walk the terrestrial Milky Way.[20] In addition to naming the star-to-star constellations that reside there, some Andean people also view the Milky Way as a negative space that houses positive (dark cloud) constellations (see Figure 2.15c). They include

- *Yutu*—the *tinamou*, a small bird resembling a partridge. There are two such constellations; one consists of what we call the Coalsack, a dark spot southeast of the Southern Cross, the other is a black cloud in Scutum
- *Hanp'atu*—the toad, another small, roundish dark area to the east of the *Yutu*

- *Machacuay*—a snake, probably an anaconda, consisting of a long black streak stretching from *Hanp'atu* to our constellation of Canis Major
- *Atoq*—the fox, a dark cloud located between the tail of Scorpio and Sagittarius
- *Unallamacha*—the baby llama, a black cloud between Alpha Centauri and Scorpio

The positions of the dark cloud constellations along the Milky Way were chosen so that their annual times of appearance and disappearance could be correlated with the biological cycles of the animals they represent. For example, the breeding period of llamas begins in late December. The animals give birth eleven months later in late November or early December of the following year and the birthing season ends in late March. Llamas usually give birth between morning twilight and noon. In the early morning hours of the beginning of the birthing period, the stars Alpha and Beta Centauri, called the eyes of the llama, undergo heliacal rise. These stars rise a little earlier each day until, by the end of the birthing season, the entire dark cloud constellation of the llama is situated well above the horizon.

By demonstrating similar correlations between astronomical cycles and the life cycles of the animals represented in the stars, anthropologist Gary Urton concludes that "the universe of the Quechuas is not composed of a series of discrete phenomena and events; instead there is a powerful synthetic principle underlying their perception of relationships between objects and events in the physical environment."[21] Dark cloud constellations are not unique to Peru. Anthropologists Claude Levi-Strauss and Gerardo Reichel-Dolmatoff have described them elsewhere in South America.[22] And similar celestial constructs have been identified by Australian aborigines and peoples of Java and Africa.

The eyes have it, but even if tropic and temperate skies were the same, what we say we see could be different, for we all make different connections, based on our varying needs and desires. How we synthesize what we see in the sky with what happens elsewhere in our lives—how we systematize our knowledge—lies at the basis of creating a calendar, one of the essential goals of watching the naked sky.

IN SEARCH OF HARMONY:
HOW TO MAKE A CALENDAR

Now that we have assembled the major events our predecessors could witness in the sky with the naked eye, we ask how these events can be used to construct a chronology. One of the basic functions of astronomy the world over lies in the development of calendars. These were devised for various reasons, ranging from the loose demands of agriculture to the more rigid dictates of a state religion. We create calendars in order to be able to predict the arrival of future events as accurately as possible—literally to reach dates in the future. To know nature's behavior in the future, we must learn from the past. People may observe the changing position of the sun at horizon, the reappearance of the thin crescent moon, the heliacal rise of a bright star or planet, the shortest length of a shadow cast by a stick, the occurrence of the first rain after a lengthy dry period, the budding of a certain plant, or the seasonal reappearance of a particular animal. All these events are concrete natural phenomena whose record of detectable patterns and interrelationships can be used as a way of deciphering what is likely to happen tomorrow. But while every calendar begins by recording and communicating a sequence of observed natural events, whether by telling myths orally or by making tick marks on the femur of a hippopotamus, it is only when two or more sets of phenomena are interrelated through a correlation that one then has a calendar.

One of the earliest examples of this sort of future-date-reaching can be found in the various attempts among cultures of the world to relate the seasonal year of the sun to the lunar synodic month, the so-called month-of-the-phases. I am going to use it as an example of how sky observers over the long run attempted to mesh cycles together to acquire a sense of order in their date-reaching schemes. This simple example can be applied to all the cycling schemes that will be discussed in our three culture areas.

As we have seen, the sun moves through a complete cycle of positions at the horizon in the course of 365.2422 days, while the moon completes its cycle of phases, from first crescent through full and new phase back to first crescent, in 29.5306 days. These periods can be determined with great precision by repeated observations made over long periods of time. For example, an observer can simply count a whole number of days between successive first appearances of the crescent moon. Sometimes the observer will count 29 days and at other times 30 days between these events. Over several years of observations, the time-averaged length of the lunar synodic month will turn out to be a little more than 29½ days.[23]

That the basic sun and moon cycles do not naturally fit together is a fact of life. The problem in making a working calendar is to invent a scheme that *makes* them fit. Now, the solar year is divisible by the synodic month twelve times with a remainder of 10.8750 days. Suppose we were to count off the months by noting the occurrence of first crescent moons beginning at an arbitrary point in the seasonal year. For simplicity, suppose further that the first of these crescents occurs exactly at the June solstice, when the sun attains its greatest northerly extreme on the horizon. If this is true, then the twelfth crescent in the lunar cycle will be recorded some eleven days before the second June solstice in the solar cycle, that is, on the 354th day of the first solar cycle; therefore, in the first solar year twelve lunar synodic months would be completed, with a little bit left over. If we were to continue the count to the second solar year we would find that the twenty-fourth crescent in the lunar series would occur about 22 days before the end of the second year (that is, before the third June solstice in the sun cycle). Again twelve full synodic months would be contained within the second solar year. But on the third solar count, the first crescent would occur on the seventh day of the solar cycle and the twelfth crescent would be recorded on or about day 331; therefore, to make things fit better, why not add a thirteenth month to that year, which would be concluded about the 362nd day of the same year, some three days before the solstice? Following this scheme, the fourth and fifth solar years would consist again of twelve months, but the sixth year would contain thirteen months, the last one ending about six days short of the solstice. Likewise, after the seventh and eighth 12-month years, the ninth year would be a 13-month year, now with a 9-day shortfall of crescent with respect to solstice.

Through discovery by observation we arrive at a practical solution to the problem of creating harmony between the fundamental solar and lunar cycles. The suggested regimen would be to insert an extra month in every third year so that the two periods will stay in rhythm. This method of inserting extra days or months into the calendar is called *intercalation*. Ideally, a calendar maker would try to devise a method of intercalation that would guarantee that the lunar and solar years would never get out of step by more than a month. It is easy to see that this simple 12-12-13-12-12-13... method can be improved by inserting an extra thirteen into the sequence once the shortfall between first crescent and solstice builds up to a full month. Ancient cultures on both sides of the Atlantic were able to develop some rather impressive intercalation formulas.[24] The leap year schedule in our own calendar is another excellent example of intercalation. It derives from attempts to fit a time period consisting of a whole number of days into a seasonal year made up of a nonwhole number of days.[25]

Did other cultures do what we did—insert those extra days? For example, did the Inca utilize a system whereby they added a thirteenth month to the calendar every year or so?[26] There is no compelling hint in the chronicles that Inca astronomers intercalated. However, there is evidence in the month lists, particularly between March and July, that some months were given two names, which can be taken to support the hypothesis of intercalation. For the Maya the evidence suggests that the 365-day year rolled on endlessly, but curiously, they did omit days from the Venus calendar in order to keep the computed mean tables of Venus we find in the surviving Maya books on track with that planet as it moved around the sky.

Creating calendars by direct observation, utilizing markers in the landscape as a check—that is, not using any counting scheme, memorized or recorded—stands in dramatic contrast to the Western scientific notion of devising a calendar by mathematical calculation. In this regard the Maya turned out to be very much like us because they wrote things down. But in the megalithic and in the Inca systems, only the architecture testifies to a precise timekeeping scheme. Observations seem to have been used only to test the theory of the calendar and to aid in tightening the correlation of its various components. By contrast, the theory that generates the numerical predictions becomes the centerpiece of our calendar; that is, the idealized model that sits on the desk, hangs on the wall, or clasps around your wrist becomes reality, and that model exists only to serve the imprecision that separates it from the facts of observation. None of us ever will recognize the first day of summer by observing the sunset on the one day of the year when it passes a vertical post implanted on a hilltop. Nor will we mark noon hour by the instant when a vertical stick casts its daily minimum-length shadow. For many ancient cultures an unwritten calendar resided in the environment itself. But it was, nonetheless, a precise calendar. Its accuracy was established through the repeatedly tested and improved observational correlation of several different sets of events involving the sun, the moon, and the stars.

Our little excursion in detecting and following unaided-eye celestial phenomena and using this information to develop calendars is cursory and by no means complete. For example, we haven't even mentioned meteorological phenomena, often neglected in astronomical considerations because modern society would tend to disregard lower atmospheric events as noncelestial. But phenomena such as rainbows, the aurora borealis, solar and lunar haloes, sundogs, tornadoes, and lightning nevertheless are mentioned in ancient historical records.[27] The main point of this chapter has been to demonstrate that nature's phenomena lie clearly in view. It only remains for us to observe carefully and record them if we wish to comprehend the sometimes complicated interrelations among events transpiring in the heavens.

The elegant rules about eclipse prediction and the meshing of planetary and lunar periods came out of an inspection of the most elementary kind of empirical data that could have been acquired by any early society possessing the time, the interest, and the means to track such phenomena once they had been spotted. The data gleaned from naked eye observation, astutely handled, yielded secrets about how the universe envisioned by specialized astronomers would behave in the future. It remains for us to determine how far a given civilization proceeded in extracting the future from the past—in making predictions from observations. This is where the cultural remains enter the picture. Equipped with the basic tools of practical sky watching and a data bank full of events to anticipate, we turn next to three culture areas of the world believed to have practiced diverse kinds of astronomy. As we begin to view the universe through their eyes, we will discover that we seem to be casting a mirror before our own.

STANDING STONES AND STARS: MEGALITHIC ASTRONOMY AND THE PEOPLE OF ANCIENT GREAT BRITAIN

. . . the circle is not only a moongate joining heaven and earth but also the hub about which the wheel of neolithic society slowly revolved.
—R. Castelden, *The People of Stonehenge*

WALKING THE STONES

As you come over the rise on the road from London onto Salisbury Plain in southern England, Stonehenge, stark and brooding, is all you can see. It appears as a unique series of rings of multi-ton worked stones accessed by a wide avenue. The archaeologists tell us that local settlers fixed the earliest grouping in place just after 3000 B.C., but the plan was modified several times in the two thousand years that followed until finally the whole place fell into disuse. Visitors always say Stonehenge looks smaller and less impressive than what they had expected—until they get close to it. If you are lucky enough to penetrate the confining fences and get into the inner-most circle (a special permit is required) you will stand in awe amidst erect carved megaliths, weighing as much as 30 tons apiece. (An elephant weighs five!) The stones seem to stare back at you like people from all directions. Those who have walked its ruins before us have dreamed up wonderful stories about why Stonehenge was built (see box).

But imagine approaching Stonehenge the way the local chieftains who built it would have gained access to it at the height of construction activity nearly five thousand years ago. You would likely have entered from the

---

THE STONEHENGE THEY DESIRED

What was the purpose of Stonehenge? There have been many answers: It was a cemetery erected by Merlin the Magician in memory of Celtic patriots fallen in war, a temple of worship built by the Romans, the place of coronation of Britain's first kings, and the house of worship of holy Druids—a native cult whose devout membership yet claims pre-Roman descent.

Seventeenth-century architect Inigo Jones, who attributed it to the Romans, saw in Stonehenge the symmetry and balance of late Renaissance architecture. (He even added an extra trilithon and warped the horseshoe into a hexagon!) Eighteenth-century Druids pronounced it the place where the supreme court of their ancestors once deliberated. This idea competed with the Stonehenge that served as an open-air planetarium. (The trilithons represented the sun, moon, and planets, the inner horseshoe's bluestones the twelve signs of the zodiac, and the Altar Stone the place where the high priest himself watched the June solstice sun rise over the Heel Stone.) One twentieth-century suggestion is that the dimensions of the Sarsen circle of Stonehenge were devised to map out the dimensions of our planet (not only its polar and mean radius but also its circumference). Another idea is that it is a landing place for UFOs (many have been sighted there), the inhabitants of which are using the ancient site to impart their superior knowledge to us. Still another Stonehenge story has the Aubrey Holes representing a count of double the menstrual cycle, the upright Heel Stone representing a penis—the sex machine of the Salisbury Plain!

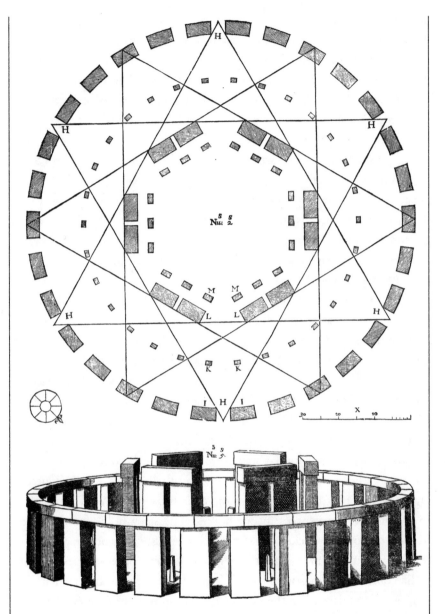

Figure 3.1 Does every age get the Stonehenge it desires? This was seventeenth-century British architect Inigo Jones's version of what the monument originally looked like.

northeast along a 12-yard-wide avenue bordered by 2-yard-high dirt banks outside of which lay equally deep ditches. That means you and six of your friends could walk abreast an arm's length apart all the way down the 2-mile-long causeway (which meanders its way toward Stonehenge from the nearby Avon River). As you walked, your tunnel vision would be confined to the ring of stones that lay straight ahead (see Figure 3.2 on page 63). Then, about 40 yards before you reached the end of the causeway, you would pass through an imposing pair of standing stones more than 15 feet high. These are the first megaliths or giant stones to be erected at Stonehenge (they were set in place about 2600 B.C.). Only the easternmost one, the Heel Stone, remains standing today. Sixteen feet of it lie above present ground level and it weighs 35 tons. It tilts like a lone sentinel over 75 yards out from the closely gathered rings of stones that lie at the center of the ruin. Next you would enter, via another pair of stone portals, the largest break in another ditch-and-bank enclosure. This one is a circle 142 yards in diameter. It too dates from about 2600 B.C.

Even though we have scarcely penetrated to the heart of the monument, questions about Stonehenge already beckon to us from the distant past. Why build in a circle? Why such megadimensions? And why of all places in the middle of a barren plain? For us, architecture has a singular function. We eat breakfast at Joe's Diner, worship in our neighborhood church, synagogue, or mosque, and spend our leisure hours at home with family and friends. But on some Monday mornings Joe's Diner becomes a place where the fans gather to rehash the weekend's football games. On Wednesday nights people play bingo games in the church basement. And some of us have computer workspaces in our homes. The same holds for the places where astronomers watched the stars. To understand ancient perspectives of the sky we must get away from the notion that their "observatories" need to be like ours—buildings with round domes filled with machinery that constantly collects data when the sky is clear. We shouldn't fall into the trap of garbing our forebears in modern scientific attire, endowing them with the same rationale we possess to philosophize, to hypothesize, and to practice experimentation.

Stonehenge is a good example of multifunctional architecture. According to some archaeologists it was a cultic center, an economic center, a place of fortified habitations, a celestial temple, and an observatory for tracking the sun and the moon—all at the same time. All of these definitions crosscut one another, some having been stressed more at one time than another in its two-thousand-year-long building history. The great achievement at Stonehenge—what makes it a wonder of the world in my opinion—is that the genius of its inventors encapsulated all of these functions in a single monument.

Around 3000 B.C., enhancing the landscape seems to have been very much in vogue among the neolithic cattleherders and pig farmers who lived in the local area. Their causewayed camps were large, crudely built structures demarcating broad hilltops. But we would be mistaken to think that what we see in the landscape today represents what *they* witnessed so long ago. There were many circular structures in the vicinity then. For example, faint remains of another earthwork lie just 3 miles north-northwest of Stonehenge, where locals dug out a roughly circular ditch 120 yards in diameter and piled the dirt around its periphery to make a bank 230 yards wide—the opposite of the Period I ditch-on-the-outside and bank-on-the-inside structure at the Stonehenge Causeway. Called Robin Hood's Ball, this structure, too, had an entryway on the northeast.

If you continued to walk into Stonehenge's inner sanctum just inside the circular bank, you would next encounter the Aubrey Holes (named after the sixteenth-century antiquarian who first uncovered them): 56 evenly spaced chalk- and rubble-filled pits between 2.5 and 6 feet wide and 2 to 3.5 feet deep, arranged in a circle just within the periphery of the ditch and bank. They are so precisely laid out that the circle that guides them surely must have been ruled out with a rope and stake placed at the monument's center. The Aubrey Holes were built about 2900 B.C.

Four large Station Stones that date from 2600 B.C. (500 years after the Heel Stone!) are arranged in a perfect rectangle circumscribed into the Aubrey circle with its axis aligning with the main avenue. A double ring of holes, 30 and 29 in number (labeled Y and Z by the archaeologists), lie within. But once you break the ring of Aubrey Holes, it is only about three dozen steps (25 yards) before you cross the next major circle at Stonehenge, an impressive 45-yard-diameter ring consisting of thirty Sarsen Stones (the term comes from Sarazen, implying foreign). These car-sized slabs stand about 16 feet high and weigh in at 25 tons apiece. Constructed about 2450 B.C., they are capped by lintels all the way around. Archaeologists think they were dragged from Overton Down, about 20 miles north of Stonehenge.

They were likely transported via sledge, with a team of oxen likely providing the pull (it would have taken several weeks per stone). There is clear evidence that the Sarsens were carefully shaped and dressed into roughly rectangular form with tenons at their tops before being cantilevered into place to receive their lithic crossframes, into which the builders had drilled matching holes.

Passing farther inward, you would next come to the double circle of eighty-five bluestones. These were also "foreign" and very different in composition (and obviously in color) from the neighboring megaliths that surround them. They were put into place about 2550 B.C., after Stonehenge had been abandoned for three centuries, and they constitute the first true circle made up of stones at Stonehenge. Whether inspired by a redevelopment project or by an invasion of imaginative foreigners who had different stylistic tastes, locals carted these volcanic blocks, possibly partway by sea, from the Preseli Mountains, 130 miles as the crow flies northwest of Stonehenge in present-day Wales. Weighing in at about 4 tons apiece, these huge vertical blocks were set up about 6 feet apart in a pair of concentric circles consisting of thirty-eight members per ring. A 6-foot-wide entryway, also on the northeast, is marked by five pairs of stones, laid out as if to preserve the old alignment tradition of the causeway and the Heel Stone gate. There is some evidence the stones were joined together, capped by wooden lintels.

It is not until you walk these last few steps into the monument's center that you encounter Stonehenge's most famous stones up close. These are the five trilithons that make up the great horseshoe-shaped arrangement clearly visible in the plan of Figure 3.2. For unknown reasons, before they completed their bluestone project, architects tore down the double bluestone ring and put the Sarsens in place. Only later did they reassemble between forty and sixty of them within the Great Sarsen Circle, along with a tiny horseshoe of nineteen more smaller stones set within the great trilithons (all this was done about 1800 B.C.). The horseshoe trilithons (so called because each is made of three stones: two vertical supports and one capstone) are made out of the longest of the bluestones. Each of the vertical supports is so close together that you cannot wriggle your body in between them. They are arranged in the shape of another horseshoe, 12 yards wide, with the open axis pointing straight out the northeast causeway. In the very center

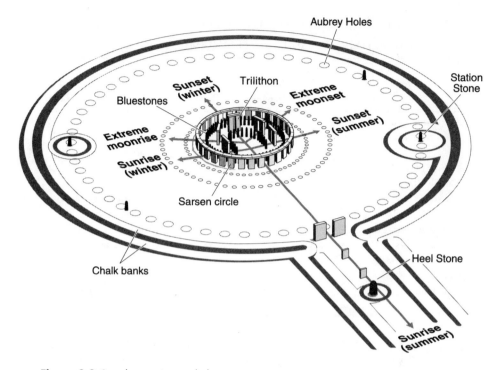

Figure 3.2 Stonehenge view and plan.

lies the Altar Stone. Now prone, it was erect when it was put in place about 2000 B.C. and may once have stood as the principal idol of worship.

So Stonehenge wasn't built in a day. In fact, it was constructed, deconstructed, and reconstructed over more than a millennium by people with rather diverse and special tastes for building materials. But before we move ahead to discuss astronomy and calendar making at Stonehenge, a reminder: At the time we took our hypothetical walk over four thousand years ago, some of these features would already have been dismantled and others would not yet have been erected. Table 3.1 summarizes the chronology of the components of Stonehenge that we will be dealing with. What we see on the ground today are the remains of many building periods, a mosaic in time of a building project that lasted more than twenty centuries. At no other time than the present has Stonehenge ever looked the way we see it now. Two stones standing only a few yards apart in space may be separated by hundreds of years in time. We need to keep this important point in mind when we discuss possible astronomical alignments at the site.

Table 3.1. The Principal Components of Stonehenge and Their Possible Astronomical Function

Principal Architectural Element	Number of Components	Building Period	Suggested Astronomical Use
Heel Stone Gate and Causeway	I	2600 B.C.	June solstice marker
Posts Adjacent to Causeway	6?	2700 B.C.	Northern lunar standstill marker
Aubrey Holes	56	2900 B.C.	Eclipse prediction computer
Double Bluestone Circle	85	2550 B.C.	?
Sarsen Circle	30	2450 B.C.	Count of days in lunar synodic month
Trilithon Horseshoe	5	2100 B.C.	Solstice and lunar standstills
Station Stones	4	2600 B.C.	Solstice and lunar standstills
Bluestone Horseshoe	19	1800 B.C.	?
Bluestone Circle	59	1800 B.C.	?
Y, Z Holes	30, 29	1800 B.C.	Count of days in lunar synodic month

When I first walked into the center of Stonehenge, my most significant question was: What would I have done after I broke through all these circles and entered the inner sanctum of the great Megalithic structure had I lived there 4,000 or 5,000 years ago? As we attempt to answer that question, we need to examine what we know about the people who contributed to building it. We will learn that many motives went into its overall design; some of them endured and others did not. But first let us come back to the present and review the modern astronomers' curious and controversial involvement in the picture.

STONEHENGE DECODED?

No one has ever doubted that Stonehenge had at least something to do with skywatching. As early as 1846 the Reverend Edward Duke, who lived near to it, suggested that Stonehenge was one member of an orrery—a planetarium consisting of seven temples that represented planets "revolving" around Silbury Hill, which signified the sun. This colossal burial mound is located about 10 miles northeast of the site. Even before that, antiquarian William Stukeley had pointed out the solstitial orientation of the great causeway we talked about in our hypothetical walk. In 1894 Sir Norman Lockyer, a respected astronomer and editor of *Nature* (he is responsible for the discovery of helium in the sun), published his *Dawn of Astronomy*, a daring book in which he argued that hosts of ancient Egyptian and other temples as well as Stonehenge were aligned with celestial bodies.[1] Stonehenge, he declared, was a sun temple (a reasonable interpretation for an astronomer who specialized in the sun) with its horseshoe axis aligned on the June solstice sunrise. He even calculated that the best date to fit the alignment with the shifting stars was 1680 B.C. Lockyer the astronomer received a harsh response from the archaeologists and historians of his day, a portent for future astronomers who would meddle on foreign academic turf.

Occasionally the scientific world drops a bombshell. In the October 26, 1963, issue of *Nature* magazine, the British-American astronomer Gerald Hawkins launched the results of what would turn out to be a very controversial study of Stonehenge. Going Lockyer one better, he suggested that the monument was an observatory designed to mark significant positions of the sun and moon as they follow their respective 1-year and 18⅔-year cycles—the horizon standstills we talked about in Chapter 2.

Hawkins's spin on the old observatory idea involved the use of a computer—then a novelty in such studies—to calculate, very accurately, the significant sun and moon horizon positions. (A whole chapter of his book is titled "The Machine.") The young astronomer demonstrated that the calculated alignments matched up with directions obtained by drawing lines between prominent megaliths on a site map (his map is reproduced in Figure 3.3). For example, using the centerpoint of the site (determined from the intersection of the diagonals between the Station Stones—STNX in the figure) as a backsight and the Heel Stone as a foresight, Hawkins discovered that the alignment between the two points coincided with sunrise on June 21 (called midsummer in Britain). This was matched, he contended, by an alignment taken between Station Stones 93 and 94, while the direction from 91 to 92 pointed to midwinter (December 21) sunset (see Figure 3.3). But then the surprise: Fourteen (out of a total of twenty-four) alignments that Hawkins discovered fit the lunar standstills. These included

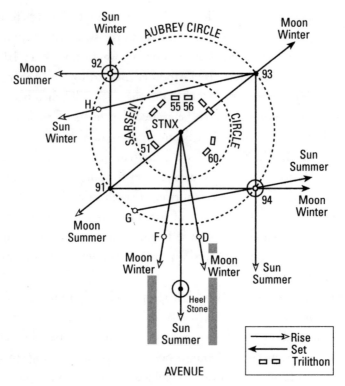

Figure 3.3 Alignments at Stonehenge viewed from above.

lines between Station Stones 91 and 94 and between 93 and 92. Other alignments taken through the trilithon and sarsen archways matched with many of these same positions.

Actually, moon alignments at Stonehenge were no Hawkins innovation. Lunar alignments at megalithic sites were first mentioned by British Admiral Boyle Somerville in 1912 but he discarded them, he says, because:

> the necessary observations of the Moon—could not have been by its rising (or setting) because, except on the day of the full Moon either the rising of the Moon, or its setting, or both, take place during daylight, and so are not visible. Orientation consequently cannot probably be connected with moonrises, except, perhaps on certain full Moon days.[2]

That ancient cultures might have kept track of time by horizon astronomy is not such a far-fetched idea, as we have seen. One advantage to astronomical horizon watching is that all it takes is a marker—a natural peak or valley or a movable stone or pair of stones (as we saw in Figure 2.7) deliberately placed in a strategic location; an observation of the sun or moon arriving at the marker constitutes a clock. No numbers, no notation are necessary. Most people who read Hawkins's paper were impressed with the deftness and skills of their distant ancestors.

Eight months later Hawkins fired off another publication.[3] This time he extended his astronomical hypothesis to envelop the fifty-six Aubrey Holes. These, he asserted, were markers that functioned as part of a digital computer used to predict eclipses. How did it work? You take three white stones, a, b, c, and set them at Aubrey Holes number 56, 38, and 19, counting clockwise from the center of the causeway. Then you take three black stones, x, y, z, and set them at holes 47, 28, and 10. By shifting each stone one position every time the sun rose over the Heel Stone as seen from the center of the site (that is, once a year) you can predict significant lunar events. For example, the midwinter full moon—the full moon nearest December 21—would rise over the Heel Stone whenever any stone is positioned at Hole 56 (every nine years). And when it does, usually there will be an eclipse of the sun or moon within fifteen days of midwinter in that year—that is, during the month containing the midwinter full moon. Hawkins's cycle keeps in step with eclipses because the time-spacing he chose—9, 9, 10, 9, 9, 10—also gives intervals of 18, 19, and 19 years, which average out to 18⅔ years for the black or the white stones. This is the same as the time it takes the moon to make one full oscillation from northerly to southerly standstill and back again, a cycle of 18.61 years which, he said, also governs eclipse frequency.

By tabulating dates of eclipses and dates and positions of moonrise, Hawkins demonstrated that the stone computer worked perfectly for about 300 years, by which time it fell behind one or two days per prediction. But ancient Stonehengers could easily correct that quite simply by intercalating—by advancing all the stones one position. Hawkins sweetened his hypothesis by providing historical evidence that Stonehenge people observed the moon, though it was all written down some three thousand years after Stonehenge's megaliths first stood erect. He cites this passage from the Roman historian Diodorus's *History of the Ancient World*, written about 50 B.C.:

> This island ... is situated in the north and is inhabited by the Hyperboreans.... And there is also on the island both a magnificent sacred precinct of Apollo and a notable temple which is adorned with many votive offerings and is spherical in shape.... They say also that the moon, as viewed from this island, appears to be but a little distance from the earth and to have upon it prominences, like those of the earth, which are visible to the eye. The account is also given that the god visits the island every nineteen years, the period in which the return of the stars to the same place in the heavens is accomplished....[4]

Fact or fiction? History or myth? Was Stonehenge the "spherical temple" Diodorus wrote about? Today, only the right-hand standing stone of the five thousand-year-old Heel Stone gate remains, but the rising midwinter full moon still shows up on time. It completes a cycle of arrivals and departures along the east horizon every nineteen years, just as the ancient historian, through the voice of the modern astronomer, seems to be telling us.

Hawkins's ideas about the scientific function of one of the great monuments of the world reached the public via his popular book, *Stonehenge Decoded*, and a documentary produced by CBS Television. Readers and viewers were fascinated. Finally the minds of our archaic ancestors had been penetrated and the revelations were startling. Stonehenge was a virtual neolithic computer and observatory built before most of Egypt's pyramids were even contemplated. Its builders came off as Bronze Age geniuses who practiced exact science the way we do. If all of this were true, our ideas about the past would be turned upside down, the course of the progress of civilization across Europe reversed—passing from west to east rather than east to west. This was the public's reaction. The professionals' retort was not so accepting.

In war, radical thrusts are often answered by strong, swift counterattacks. So, too, in academia. British archaeologist Richard Atkinson, who had spent much of his professional life excavating Stonehenge and had written books of his own on it, wrote the first scathing review of Hawkins's book. It appeared in the British archaeological journal *Antiquity*.[5] Amusingly titling his work "Moonshine on Stonehenge," the skeptical excavator poured invective over Hawkins and cold water over many of the young scientist's ideas. Calling Hawkins's work uncritical, tendentious, and beyond the bounds of logic, Atkinson challenged the astronomer's general orientation to problem solving, his accuracy and use of statistics, his misuse of the chronology acquired from archeological data, and his whole interpretation of history. If you demonstrate that something *can* be done, this does not mean that in fact it *was* done, argued Hawkins's archcritic. For example, if someone sets sail across the ocean on a raft made of reeds, this constitutes no proof that ancient races once migrated from Africa to Yucatán or from Japan to Peru. Such a criticism flew in the face of Hawkins's underlying assumption, stated in the preface of his book, that "If I can see any alignments, general relationships or use for the various parts of Stonehenge, then these facts were also known to the builders."[6]

Atkinson further claimed that many of the so-called matches between stone alignments and astronomical events at the horizon were purely accidental. Using a computer, Hawkins argued that the probability that the positions he chose are marked by chance alignment is less than one in a million. Using a pencil and paper, Atkinson claimed instead that the odds are close to 50 percent, largely because of the huge differences Hawkins allowed when matching stone alignments with astronomical directions. (Try it yourself. See Appendix A.) Apparently Hawkins never measured alignments at the site itself. Atkinson was further concerned that the map Hawkins used to acquire his information, a map published by Atkinson and used without acknowledgment, was not the most accurate plan available and was certainly not suitable for precise work. Finally, the selection of stones used in the alignment study seems to have been made without regard to the fact that they were put up, taken down, and repositioned at different times over a 2,000-year period.

What bothered Atkinson more than all of the procedural complaints was that what Hawkins contended about the astronomical knowledge of Bronze Age society did not square with what we know of the people who built the great structure, who were regarded as barbarian. Why, asked Atkinson, would a simple farming and hunting community care about building computers to precisely chart the course of the sun and the moon? Could anyone even have been able to follow the course of celestial objects in the cloud-bound environment of Great Britain? This would have been a full-time job requiring careful observing and staking out of sightlines over many lifetimes. And without a system of writing, how could one generation pass on the relevant information they had acquired about cycles to their descendants? All this precision just did not fit with the prevailing view of British prehistory in the mid-1960s, which yielded up no other evidence to support the idea that Stonehengers were more advanced than their contemporaries in the Middle East. In fact, they were probably a lot less advanced. At the time the Stonehengers were erecting their first ditch-and-bank structure, the Fertile Crescent already had boasted the first of its great walled cities. Uruk (the biblical Erech mentioned in Genesis 10:10), spread out over 300 acres and housed 10,000 people. Its monumental buildings, advanced irrigation techniques, organized labor, copper tools, and the beginnings of a system of writing on clay tablets (cuneiform) give testament to a highly civilized urban environment.

An equally skeptical response to Hawkins's ideas came from British historian Jacquetta Hawkes, who offered a much-needed historical perspective. The clever title of her review, which also appeared in *Antiquity*, rivaled Atkinson's: "God in the Machine." It is in the opening sentence of this review that we find the oft-quoted statement that "Every age has the Stonehenge it deserves—or desires."[7] She reminds us (as did antiquarian Horace Walpole nearly two centuries before) that those who look at Stonehenge always seem to endow it with whatever kind of antiquity he or she is particularly fond of at the moment. (Recall the box on page 58!) For example, a handful of members of the Ancient Order of Druids, an old Celtic caste (actually it was founded in 1781), still shows up every Midsummer's Day for an initiation ceremony in which novice priests dressed in white robes parade their way down the Heel Stone causeway and into the magic circle. They claim Stonehenge as their own, a temple of their ancestors. Thousands of New Age followers also show up on the first day of summer to convene in a cosmic happening. They look for meaningful ties to a sacred past they imagine to have been more serene than the present (they do the same at some of ancient Mexico's pyramids—see the "serpent phenomenon" on p. 145). And when Inigo Jones, King James I's surveyor, excavated the center of Stonehenge in the 1620s, he called it a Roman temple and reconstructed it in plans and drawings so that it looked perfectly symmetrical, the way any well-educated Renaissance mind would conceive of most great works of antiquity. By imagining six trilithons (instead of five) in a hexagon about the center, he created the Stonehenge he desired. (See Figure 3.1.) We will see later that the Spanish chroniclers exhibited similar views of sites they visited in the New World.

Scientists often take a different view of history than archaeologists and historians. Believing firmly in the idea of progress, they tend to value in the past what they see as the underpinning of their own modern scientific culture—Aristotle's logic, Galileo's instrumentation, Newton's abstract laws. Often they tend to disregard context—Aristotle's belief in an animate universe, Galileo's or Newton's ideas about God. So it is not surprising that much of the scientific community responded positively to Hawkins's bombshell. He was giving them the Stonehenge they desired by portraying their prehistoric ancestors to be as astronomically oriented as they were. Thus, British astronomer Sir Fred Hoyle immediately agreed with his colleague that Britain's most famous ancient monument was a computer, but he went one step further. He called Stonehenge an *analog* computer (a model that duplicates the sun's and moon's movements), with the ring of Aubrey Holes representing the ecliptic and the hypothetical Marker Stones the sun and moon moving about it. Still, critics inquired, if fifty-six (the number of Aubrey Holes) was such an important number, then why isn't it displayed in arrangements of stones at other megalithic sites? And is the 56-year cycle real? Does the increasing complexity of astronomical observations and ideas fit with the archaeologists' findings about the different building periods? Table 3.1, which lists all the major features of Stonehenge along with their building times, indicates that the Aubrey Holes were dug relatively early in the building of Stonehenge. Yet, if we believe Hawkins and Hoyle, they constitute the most advanced stage of megalithic astronomy: eclipse predicting. Moreover, the Station Stones, presumably used to set up the alignments that would lead to the Stonehengers' ability to compute eclipse cycles, would have lain on top of and obliterated some of the holes. Finally, for the eclipse-computer theory to hold water, we would need to believe that the basic guiding principle, that Stonehenge should be an observatory, endured for more than twenty centuries. And so, the Stonehenge controversy raged on through the 1970s. Meanwhile other astronomically aligned sites in the British Isles were beginning to come to the public's attention.

Far to the north at Callanish, on the Isle of Lewis in the Scottish Hebrides, lies a site even more spectacular than Stonehenge, at least as far as astronomy is concerned. There, Hawkins discovered that when the moon migrates to its 19-year southern limit, it just barely rolls above the south horizon. It perches momentarily in a notch, then it glides forward and downward on a low-angled course back into the earth. Whether this was the moment that timed a lunar observation four millennia ago we cannot say for sure, though the view from the center of the enclosed ring of standing stones on the appropriate occasion is truly impressive (Figure 3.4).

Hawkins's work also brought to light other ongoing investigations of neolithic mensuration and astronomy. Since the 1930s Alexander Thom, a Scottish engineer at Oxford University, had been conducting studies at dozens of megalithic sites throughout the British Isles. Thom had proposed the idea that all the sites were laid out with a standard unit of measurement, the Megalithic Yard (2.72 feet or 0.82 meter) and that they possessed six different round shapes, ranging from perfect circles to flattened eggs. Like Hawkins, he had also argued that many of them were functioning astronomical observatories, their component standing stones aligning with conspicuous bumps and notches on the horizon that marked the standstill positions of the sun and the moon. Interestingly, Thom got a better reception than his more assertive young colleague.

Figure 3.4 The sun imitates the lunar standstill at Callanish.

What seemed appealing about Thom's theory to Atkinson and other archaeologists was that it was based on meticulous on-site measurements. It also began to fit with the newly cultivated 1970s image of Neolithic people as far more skilled and civilized than had hitherto been believed. The archaeological record confirms that by the middle of the fourth millennium B.C., people were already clearing the land for cultivation of cereals and for pasturing livestock. They had domesticated dogs for hunting and shepherding, and they lived in small communities of timber, sod, or stone houses 13 to 16 feet across, which they furnished with ovens, beds, tables, chairs, ceramic wares, and a variety of household tools.

But were there among them architects so well organized, so sophisticated, that they could develop their own "Bureau of Standards," with a fixed rod for laying out all their architecture along with an "Institute for Advanced Study" for inquiring into the mysteries of complex geometry? According to Sir Fred Hoyle, anyone capable of such achievement would need to possess a level of knowledge and skills far higher than what we might expect from a loosely organized community of farmers. While this prospect posed no problem for Hoyle and the scientists, it stretched the credibility of both archaeologists and prehistorians, who interpreted the evidence they had gathered to imply that Neolithic and Bronze Age people had been neither organized regionally nor subject to the rule of any "nationwide" central authority.

Inspired by the work of Hawkins and Thom, a whole generation of archaeoastronomers, most of them Americans and many of them trained scientists and engineers, took to the field to explore their favorite ruins in search of astronomically significant alignments. They ranged over the Cahokia Mounds near St. Louis, Missouri, to the Medicine Wheels of the far western United States and southwestern Canada, to Puebloan sites in the Southwest, and down to the Mexican pyramids. The Big Horn Medicine Wheel, located in the mountains of Wyoming, became North America's Stonehenge. Its arrangement of 12-yard-long spokes made of piles of boulders radiates outward, ending in cairns that align with the summer solstice sunrise, as well as with the rising positions of stars whose heliacal risings timed the lunar months when rites might have been conducted there. (Later studies have called into question much of the astronomical argument.) In Mexico, alignments to celestial bodies were sought for peculiarly oriented cities, like Teotihuacan, as well as oddly shaped, skewed buildings like Structure J at Monte Albán (see Figure 3.5). Two archaeoastronomy journals sprouted up in which

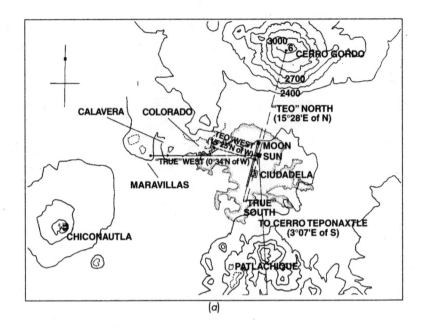

(a)

(b)

Figure 3.5 Astronomical alignments in Mesoamerica. (a, b) Teotihuacan is a city skewed to the cosmos, the place where the Aztecs, their cultural descendants, say time began. The east-west axis, marked by a pair of pecked quartered circles, points to the setting position of the Pleiades star group, which rises heliacally on the same day the sun passes the zenith. Other markers (connected by lines in the figure) also figure in the plan. View in the photo is from the north overlooking the sun pyramid at left and the north-south Avenue of the Dead at center. (c, d) Monte Albán—another sun-star timer. A perpendicular line extending from Building J at Monte Albán indicates the point on the horizon where the first-magnitude star Capella rises. On the one day of the year when Capella undergoes helia-cal rise, the sun passes through the zenith. This event can be viewed through a vertical sight tube in Building P (dotted line). In the photo, the two buildings, viewed from the south-west, are clearly visible and the horizon point is indicated.

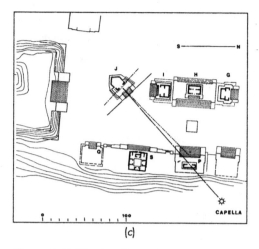

(c)

(d)

Figure 3.5 (Continued).

investigators could report the results of their work; numerous conferences were held. The "Thom paradigm," as I have called it,[8] was in full swing.

By demonstrating that precise quantifiable knowledge about the sky could be gleaned from alignments measured from a site map (or better, obtained even more exactly in the field), Hawkins and Thom touched off a tidal wave of investigation of ancient remains all over the world. What appealed most of all to this new generation of investigators was the idea that ancient people could set up an unwritten record of astronomical knowledge. The procedure for tapping into the Thom paradigm went like this: First, look for the solstices and if you're successful, then seek out the lunar standstills. Sometimes valuable insights about basic site orientation patterns were achieved,[9] but astronomically trained scholars, working alone and with little familiarity with the archaeological literature, often overlooked important data on cultural remains. Discovering an astronomical alignment, even with great rigor, is not enough. One needed to ask: What was the alignment used for? How does it fit with what we know about the society that built it? Is it part of a set of changing customs acquired by the society, like using celestial timings to regulate a newly developed system of irrigation? When the cultural record is scant, as it is at the British megalithic sites and at many of the North American ruins, we stand little chance of learning much about the actual practice of astronomy.

Despite all these problems and reservations, by the 1990s we began to acquire a much clearer picture of what the people who built neolithic-Bronze Age Stonehenge were like, where and how they lived, and what activities occupied them. Archaeoastronomical investigations have taken on a more interdisciplinary character. As a result, we have begun to find a new place for astronomical studies in the ancient megalithic world. It is based more closely upon what we now know of the people who erected the great monument.

THE PEOPLE WHO BUILT STONEHENGE

Was it the same desire held in the hearts of Egyptian pharaohs who built the pyramids, medieval worshippers who erected Gothic cathedrals, or even early twentieth-century city dwellers who fashioned skyscrapers: the idea of immortalizing themselves in the permanence of great architecture? What sort of megalomania must the Stonehengers who eventually chose to build their works of bigger and bigger stones have possessed?

When populations increase, as they did in northern Europe after the end of the last Ice Age about 12,000 years ago, people begin seeing more of one another. They bond together, they conflict, they share, and they feud. Sometimes their most grandiose communal building enterprises stick up out of the landscape like the tips of so many icebergs, leaving us to wonder what lies beneath the structure of the social fabric of the people who created them. Ditch-and-bank structures, round structures, henges (structures consisting of an external bank and internal ditch made into the shape of a circle penetrated by one or more entrances)—all are architectural concepts known originally to have been executed in wood and employed by the native population of neolithic Britain as everyday living spaces by the mid-fourth millennium B.C. Archaeologists have named the earliest people on the Stonehenge scene the Windmill Hill culture, after the type site, a causewayed camp some 20 acres in extent. Their ancestors crossed the English Channel a thousand years before Stonehenge was built, bringing with them to the island its first sheep and goats. They settled on the chalk downs of south England, domesticated pigs and cattle out of the wild (archaeologists have found garden shovels made from the shoulder blades of cows), lived in sod houses, and got into the habit of burying their dead in round mounds.

Communal timber roundhouses that housed small extended families made their first appearance in the third millennium B.C. The earliest of them measured just a few yards across. Their walls often were made out of wattle and daub (twigs held together by clay or mud), and their cone-shaped timber-frame roofs were covered with thatch. Sometimes a few dozen or more people lived in larger lodges that were subdivided into separate rooms that housed different families connected to one another by bloodline. Durrington Walls (Figure 3.6a), just a few miles northeast of Stonehenge, was among the largest living units. Here was a whole town in the round, a ditch-and-bank structure 525 yards in diameter penetrated by a pair of axially arranged entrances and consisting of a dozen or so round buildings in each of which several families lived. Woodhenge (Figure 3.6b), also nearby, at one stage measured 48 yards across. Functioning as a place for both communal living and worship, like Stonehenge it featured a ditch-and-bank structure and an accessway directed to the northeast (it was built about 2300 B.C.). The central area of Woodhenge, which likely housed totem poles, was (like Stonehenge) open to the sky. The remains of a three-year-old child at the dead center of the Woodhenge complex hint at the practice of human sacrifice, perhaps a foundation ritual.

By the middle of the third millennium B.C., the so-called Beaker culture (named after a type of pot found in the round barrows) who settled the area were well organized into little chiefdoms. Having migrated from central Europe, they used the bow and arrow and metal daggers, which gave them an enormous advantage over indigenous locals with whom they eventually merged to form a new culture.

What was Woodhenge like? Geographer Rodney Castelden speculates: "It must have been the epitome of the great mother forest beyond the town gate, an ordered microcosm of the natural world, with a symbolic clearing at the center letting in the slanting shafts of sunlight."[10] But why the ambiguous ditch-and-bank structure? Protection comes to mind, but is this interpretation just a projection of our own culture's defense-mindedness? True, circles may appear to offer greater protection than any other geometric form, but maybe they were thought to possess some magical power, too. Perhaps the circle was as much a device for sealing in the sanctity of the community within as a means of thwarting invasion by an enemy from without. We will

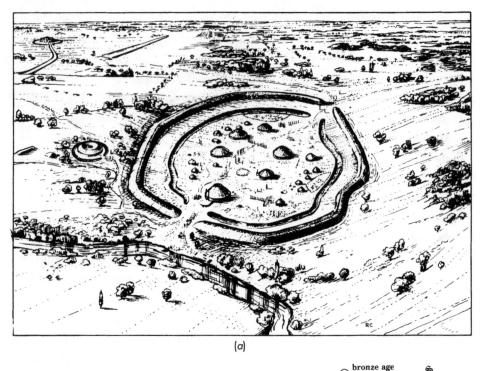

(a)

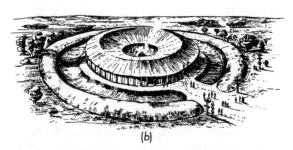

(b)

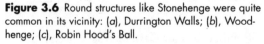

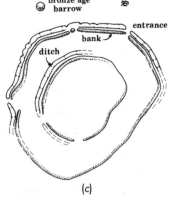

(c)

Figure 3.6 Round structures like Stonehenge were quite common in its vicinity: (a), Durrington Walls; (b), Woodhenge; (c), Robin Hood's Ball.

never know why, but more than most ancient cultures these people seemed motivated to build in the round. And Stonehenge, despite the huge megaliths at the center that seem to attract all the attention, is basically an earth circle.

Large henges may have served as communal gathering places in which local families or clans assembled, made agreements, talked politics, prayed, discussed news, and traded goods—in short, places where people acquired their social identity. To make such important places more enduring, locals adopted the style of building their central places of worship, their tombs, and their circles in stone. They extended building concepts acquired from everyday domestic life, yet still employed timbers in the construction.[11] Archaeologist Aubrey Burl believes that the tradition of the woodworking capacity of these ancient people was embodied in the linteled structure of the Sarsen Ring and trilithons, and in the vertical arrangement of standing stones that resemble the timber-settings in the domestic roundhouse. Twentieth-century archaeologists have determined that many of the holes and pits on the Stonehenge map were probably postholes that once may have constituted walled structures, so the idea for the round plan of Stonehenge is not so unusual. It derived from the conversion of a communal wooden roundhouse into a more permanent ceremonial center. The symbol of the family, the domestic dwelling place, became architecturally fixed in the ring of very large, round, standing boulders we call Sarsens.

Though the earliest evidence for a causeway penetrating an enclosure comes from about 3600 B.C., in larger structures like Stonehenge the causeway may have been more than just a means of access. Rites may have accompanied the access to the inner circle by members of the community. The Stonehenge accessway may have been the penetrating strand from the outside world by which people sealed their social ties to brethren in the private circle within.

The archaeological record also tells us that initially Stonehenge must have been erected by several extended families, each consisting of approximately fifty to one hundred people. They were probably colonists who came in search of a good place to cultivate grain and later to raise cattle. There is evidence that their descendants and other newcomers may have mismanaged the environment to the point of crisis, denuding it of trees and offsetting the chemical balance in the soil, a conclusion that resonates all too readily with some of our own contemporary ecological concerns.

By 2500 B.C., the population had increased, tending more toward pastoralism, intensive farming, and a little mining. This is when they lived in huge roundhouses erected upon the high moors. Scholars agree that the people who constructed the massive trilithons must have been of high rank, with specialized groups assigned individual roles in the building project. Archaeologists call this culture the Wessex culture, after an excavation in southern England that reveals a chief-led society that was in contact with similar cultures on the Continent. They would need to have been specialized to set up the great trilithons. After transporting the worked 25-ton slabs on rollers to the site and positioning their bases over a predug ramped pit, they probably raised them with levers, a step at a time, until the bottom of each slab slid into the hole. Then they lifted it upright by pulling horizontally on ropes attached to a wooden frame atop the slab. (Between 100 and 200 men could have done the job.) They capped each trilithon by elevating its 6-ton lintel on a platform snug up against the two uprights. They raised the lintel by placing packing material under each end and elevating it in steps until it could be rolled into place and its sockets fit into plugs carved at the tops of the verticals.

What were the ancient Stonehengers like at the height of its grandeur? When the celebrated "Ice Man" was accidentally discovered in the Alps in the early 1990s, after having been hidden under a recently melted glacial cap for over 5,000 years, the public was fascinated most of all by the fact that he seemed so much like us. He had stuffed his shoes with straw for insulation, he wore socks and underclothing, and he carried a quiver and a little pack containing his belongings. The Ice Man was tattooed and well groomed; his hair had recently been cut. Perhaps because his appearance and habits seem so familiar to us we empathize with this unfortunate human being. Could the poor soul have lost his way on a visitation or a hunting expedition? Had he been in exile? Was someone pursuing him? It is almost a cliché for archaeologists to remind their readers that people of the Stone Age had skulls and brains the same size as ours; but, symptomatic of the progressive age, many persist in believing our ancient ancestors were howling savages teetering several steps down the ladder of social development from us—except when occasional finds like the Ice Man or the exquisite 30,000-year-old French cave paintings of Cosquer and Chauvet remind us of the sophistication of our forebears.

Remains indicate a typical Stonehenge male stood about 5 feet 7 inches tall, and a typical female 5 feet 2 inches. Like most people who lead active lives and partake in a low-fat diet, these people were fairly slender. Their faces were relatively long and they had small, slightly turned-up noses. By our standards theirs would have been a very youthful society. According to Aubrey Burl, most of them died young—in their twenties—which means that roughly half the population would have been younger than twenty. Cancer and heart disease, along with other stress-related maladies that plague contemporary society, were likely uncommon. Life-threatening disabilities would have included arthritis, tetanus, tuberculosis, polio, sinus problems, and for women, childbirth. They may have acquired animal-related diseases like anthrax from living in close quarters with their cattle. We know that only a small percentage had tooth decay, and their whole-grain, largely vegetarian diet was likely superior to that of the modern world. Periodic famine caused by uncontrollable drought and flood would have placed the population at the mercy of the climate to a much greater extent than today.

We don't know what these people wore, for not a shred of cloth survives beneath damp British soil; but based on excavations of early Bronze Age burials in Denmark, Rodney Castleden has put together this intriguing description of a teenage girl from Egtved who, he thinks, would not have been too different from a Stonehenger:

> She wore a miniskirt 65 centimeters [about 25.5 inches] long made from vertical woollen strands that were gathered at the waist and hem in elaborate edgings; the waist edging was tied in a bow below her navel, with the loops hanging down in front. The skirt was slung low on her hips, so that her stomach was exposed. She also wore a short brown woollen tunic made in one piece, with gussets at the armpits and sleeves reaching just below her elbows. The neck line was high, wide and hemmed. On her stomach she wore a circular ornamental disc mounted on a woven belt with a large tassel on one end. The belt was 2 meters [about 6 feet] long and was wound round her slender waist several times.
>
> The presence of flowers in her hair suggests that she was in summer dress and we would hope that she wore something warmer in winter.[12]

Coiffures included piling the hair atop the front of the head, perhaps adding a false hair ornament on top of that for stature. Women carried combs and wore hairnets and bonnets. They even slung decorated handbags about their belts.

The men were often clean-shaven and also wore caps to cover their long hair, which frequently was parted in the middle.

Castleden writes:

The man's basic garment was a deceptively simple tunic that wrapped round the body from shoulder level down to knee or mid-calf. He fastened it round his waist with a leather belt and over each shoulder with a leather strap. The cut of the breast-line varied: it could be horizontal, or sloping down to one side, or tongued up to the throat. Over this tunic he wore a knee-length woollen cape that could be round, oval or kidney-shaped; it was fastened across his chest and the edge was flipped back at the neck and chest to form a collar.[13]

Pendants often were hung about the neck. Both genders wore moccasins made of leather or simply a piece of cloth bound around the feet and ankles and both decorated their faces with rouge. They also colored their lips and probably painted decorations on their bodies.

Vignettes of the social life and personal habits of Neolithic people based on an examination of their remains lead to an impression of the average Stonehenger as far more sophisticated than was thought just a few decades ago. But was there room in their lives for a science as esoteric as astronomy? And if so, what sort of skywatching would fit their lifestyles? How would astral knowledge be used? What needs would it satisfy? These are the questions that intrigue archaeoastronomers the most.

ASTRONOMY
AND STONEHENGE SOCIETY

Why should ancient Bronze Age people have cared about what happened in the sky? As I stated earlier, the cyclic movement of the sun, moon, planets, and stars represents a kind of perfection unattainable by mortals. The regular occurrence of sunrise and· moonset could have provided the ancients with a dependable and orderly sense of time, a stable pillar to which they could anchor their thought and behavior. For example, they could easily have followed the sun wherever it went, marking its appearance and disappearance with great care. The sky god's return to a certain place on the horizon would have constituted one of many signs in nature that indicated the time was right to plant the crops, that the nearby river would soon overflow its banks, or that the monsoon season was about to arrive. The planting and harvesting of crops could have been regulated by a whole series of celestial events, many of which were discussed in the preceding chapter.

We cannot overestimate the importance of predicting and following seasonal change among these people. For them, time was activity itself. It was *lived* rather than kept on a watch. Getting a sense of time from skywatching is difficult for us to appreciate because we no longer have need of practical astronomy in our daily lives; therefore we tend to pay little attention to the heavens, except maybe as an occasional diversion. We might catch a sunset, glimpse the Man in the Moon, or notice the evening star. But we would never think of such casual observations when setting up appointments or scheduling activities. The artificial clocks by which we pace our daily activities give us a distorted view of the dependence of human time on phenomena that happen in nature and in the heavens.

There was more to prehistoric astronomy than timekeeping. Skywatching influenced many different aspects of ancient cultures. We find the sun, moon, and stars woven into myth, religion, and astrology. Representations of deified celestial luminaries adorned their temples as objects of worship and they were symbolized in sculpture and other works of art. Believing they lived in an animate universe made up of component sky and earth deities who were in one way or another extensions of themselves—fighting endless wars, loving and hating, surviving life after death—many of our ancestors who followed the time-bearing luminaries attached rites or reenactments of life's vital functions to special days on nature's calendar. They would hold feasts and make offerings in order to pay their debts to the gods for a bountiful year or perhaps in anticipation of a better crop yield after hard times. The ancients literally would talk to the sun and moon, converse with the planets.

Given this scenario, staking out a set of markers in the landscape to graph the whereabouts of the gods would have been part of ancient common sense. (Try it in the example I give in Appendix A.) Devices that could predict sky phenomena well in advance would offer one group of people a powerful advantage over their uninformed neighbors. The space within which they followed the sky gods and conducted their discourse with them might well be regarded as hallowed ground, a sacred space administered by the wisest of the wise.

You don't need mathematics or writing, much less telescopes and computers, to practice this kind of astronomy; better we should explore how the sky is tied to a people's worldview, how it relates to their religion, their politics, and their economics. To do so we need to get away from imposing our own beliefs about nature upon them. Above all, we need to be careful about separating the natural world from everyday behavior as we have done in our own culture, lest we stereotype all astronomers, past and present, as quantitative scientists who use high-tech equipment in order to test their hypotheses about how nature behaves.

Based on the material record they have left behind for us to probe and ponder, how might the builders of Stonehenge have made use of the sky as a way of finding meaning in the world they confronted? How might they have used what they saw to order and structure their beliefs about the world around them? And, most importantly, how does the evidence we find in the Stonehenge alignments fit their worldview?

The first impression I had of Stonehenge when I walked around it and then stood at its center was that it was a sanctuary, a place of assembly, a space for people—a very special place, fashioned on a grand scale out of diverse precious materials like bluestones and Sarsen boulders. Whatever Stonehenge was, it was extraordinary and awe-inspiring to me. Architectural historian E. C. Fernie has compared the trilithon horseshoe and its accessway with the ambulatory and apse of medieval churches such as the Canterbury Cathedral.[14] Rodney Castleden sees it as the earliest of the magic circles, the construction of which he views as a continuous tradition traceable in Europe all the way up to the present.[15] Mystics believe that you can magically protect yourself by drawing a circle around the place where you stand. In medieval times the magician did it to conjure up spirits or to protect himself from the devil. To step outside, even to break the circumference with an arm or a leg, was to court disaster. He frequently used a peg and a rope to rule a pair of concentric circles, the ring in between serving as a space to write down the magic formulas that facilitated contact with the spirit world. Other magicians carried with them folded paper circles, which could be set up at a moment's notice. Two thousand years earlier, Etruscan architects built great pre-Roman cities, but not before bringing a diviner to find the most auspicious site. He slew a sheep, removed its liver, and gazing into its shining red consciousness, literally read the minds of the gods. Once the site was selected he consecrated the work by plowing an encircling protective furrow around the area where they would build the city. One thousand years before that, about the time the Sarsen circle went up at Stonehenge, the Sumerians wrote prayers on clay tablets devoted to circles of protection.

Did the Stonehengers merely witness sunrise or did they believe they made sunrise happen? Did the animated forces of the universe feast, perform, and dance with the people of Stonehenge within the magic circle? Were the stones themselves perceived as animate beings, just like the people who put them there? Did the participants dance with the upright megaliths, encouraging the sky gods on their way, preparing to welcome them when the dazzling celestial lights came to the appropriate designated spot? Could Stonehenge's circular design, repeated over and over for two thousand years, be a reference to the shape of the sun disk upon which they depended for food, warmth, and life? Stonehenge was always open to the sky, and while we can argue about the slightly eccentric position of the Heel Stone,[16] its main avenue more or less does target the northerly extreme of the rising sun. When the sun god finally kept his annual appointment, just as his worshippers had arrived at the magic circle, the people would have been offered an impressive view as its rising rosy radiance glinted off the top of the Heel Stone. A pair of upright slabs situated at the inner portal (at the closure of the circular ditch and bank in Figure 3.2) would have framed the sunrise and made it look even more impressive.

I am convinced that if Stonehenge had anything to do with sun and moon astronomy, the association between its architecture and the sky was more closely allied with theater than with exact science. I think Stonehenge was built to celebrate the entry of the sun god into the circular sanctuary. It was designed to chart his course as well as that of his rival, the one with the silvered, oft-slivered variable countenance, who migrated even farther to the northern and southern climes. In southern Great Britain the June solstice would have been the time to look forward to a bountiful crop; the sun would then follow its highest track as it slanted across the sky, officially initiating the peak of the growing season.

Once the northerly summer solstice was recognized, why mark out the southerly winter one? Here is where the other alignments of the Station Stones and great trilithon archways may fit in. I do not believe they were laid out simply for reasons of symmetry. Targeting where the sun god migrated

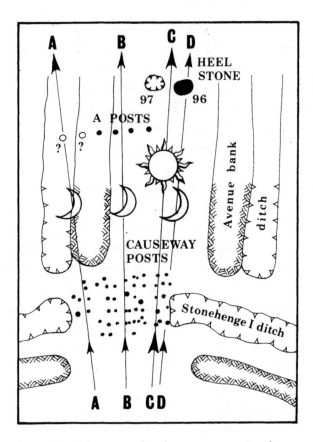

Figure 3.7 Enlargement of northeast entrance to Stonehenge (see Fig. 3.2) showing stones and posts that may have charted solar as well as lunar standstills. (A) Northern lunar standstill, (C) midsummer sunrise, (D) halfway point of lunar cycle, (B) one-third of the way.

in winter may have been an even more important act than demarcating its summer excursion limit. This would have marked the most barren season of the year, when perhaps more than in any other part of the annual cycle people sought rapport with their life-giving deity. In fact, the view on this day is just as impressive as its summertime counterpoint, for if you walk the accessway into the center of the circle on midwinter sunset day, you face the sun and you watch it become enveloped within the trilithon horseshoe. Now would be the time to plead with the sun to return, for at this season of the year he had already spent too much time in the lower world. Perhaps the god will leave us forever. Should we uproot ourselves to follow?

Where would the moon logically fit into this picture? By searching for eclipsed full moons, Gerald Hawkins may have looked past a more fundamental aspect of lunar observation that builds sensibly on the Stonehenge alignments directed to the sun's horizon extremes. In addition to the bright light it provides, the full moon also mimics the sun in mirror image. It passes high in the sky around the time of the winter solstice and lies low in the sky in summer. As the old Babylonian creation myth says, you always see the full moon rising directly in front of you just as the sun disappears behind you. For a short time around sunset, both sun and moon may be visible on opposite horizons. Seeing is believing; little wonder that in cultures all over the world the sun and moon make up the complementary halves of a cosmic duality.

On the night of the full moon, the lunar god or goddess takes over the duties of lighting up the world. Just as daytime's great illuminator vanishes from view, the lunar disk rises, glowing ever brighter with each degree of ascent. This is the one day of the month that bears light from dawn to dusk and back to dawn again. Midwinter full moonrise at Stonehenge may be even more dazzling than midsummer sunrise. It always happens within 10° of the Heel Stone and it occurs on the same day the sun sets in the trilithon archway. The count of the Y and Z holes located just outside the Sarsen circle, 30 and 29 in number, respectively, may have marked out— again cleverly in the round—the days of the lunar month (see the discussion on how to determine the length of the lunar synodic month on p. 31). When the count of these holes—commenced at the opening—returned back to the opening to the inner circle, another full moon would be scheduled to appear.

Did the builders of Stonehenge go further? Did they recognize the lunar standstills and the 19-year cycle of the moon? If so, did they discover a way to use this period to predict eclipses and even devise a computer for generating those predictions? To answer these questions we need to reassemble the evidence. If the answer is yes, we deserve to be surprised by the extraordinary capabilities of these early astronomers.

Recall the alignment evidence. As Hawkins showed, there are plenty of moon alignments at Stonehenge to support the idea that the builders could have been interested in tracking the moon over long periods of time. We also know there were the "A posts" in Figure 3.7 adjacent to the Heel

Stone that once held wooden posts; these, too, may have marked the progress of the moon toward its northern 19-year standstill. Within the ditch-and-bank lay the four Station Stones. Alignments among these stones marked out the maximal stretch along the horizon between the sun at one of its standstills and the moon at its opposite standstill. As Hawkins has pointed out, a 90° difference between these alignment pairs occurs only close to the latitude of Stonehenge, so we can think of the Station Stone rectangle as fairly unique to sites located within, say, a 60-mile-wide band running east to west across Great Britain. Despite these facts, I find it unlikely that a small group of people intent on following the moon would have roamed the moors of south Britain to find the correct place to build their "magic rectangle."

Are these alignments accurate enough to lead to eclipse prediction? What look like time markers to us may have been used simply to frame moonrises the way the Heel Stone gate framed sunrises to make them look more impressive, rather than functioning to predict where the moon would pass in the future. In fact, lunar limits may not even be relevant to eclipse prediction. Recall that the arrival of the moon at a standstill is based on the time it takes one of its nodes to return to the same spot on the ecliptic (what modern astronomers call the period of regression of the lunar nodes).[17] Now, in relation to horizon observations, this period returns the moon to the same extreme point of the horizon. But because it is not an exact multiple of the draconic month, which is a necessary condition for setting up eclipse cycles (see Table 2.1), eclipses of the full moon need not and probably will not occur at the same time in successive cycles.

The lunar standstill period of 18⅔ years (or 6,797 days to be precise) that Hawkins investigated is often confused with the Metonic cycle, a 6,940-day cycle named after the fifth-century B.C. Greek astronomer, Meton, who discovered it. Recall that the latter cycle marks the period in which a given phase of the moon will recur on the same date of the year. Because the Metonic cycle, unlike the lunar standstill period, *is* a whole multiple of both the draconic and synodic lunar intervals,[18] a lunar eclipse is likely to occur on the same date as an eclipse during the previous cycle.

Is the case for Stonehenge as an eclipse predictor built out of a house of cards? In *Stonehenge Decoded,* Hawkins offered little justification for including the lunar horizon standstills in the Stonehenge alignment tests along with the solar extremes. He appears to have sought them out as a logical next step once he discovered that Stonehenge marked the solar horizon limits. Apparently, once he decided that he had achieved a positive lunar alignment correlation, he began to analyze the Aubrey Holes as a seasonal eclipse counter; it is only then that he proposed the eclipse-observatory hypothesis for Stonehenge, evidently believing that the lunar limits served as a tool for predicting eclipses.

Recall that Hawkins also considered it possible that Diodorus's statement about a temple on the island of the Hyperboreans may have referred to Stonehenge and that the 19-year period mentioned in the statement could be connected with the lunar standstills and the nodal regression period. But the last phrase of the quotation from Diodorus on p. 68 clearly identifies the 19-year period with the Metonic cycle and not with the nodal regression cycle. Alexander Thom also justified the search for ancient observations of the lunar standstills, but for a different set of reasons. He inferred that "men living on the coasts of the western ocean must have noticed the connection between the tides and the Moon."[19] He claimed that this was a major factor that compelled Megalithic people to study the movements of the moon. Eventually, the skywatchers of Stonehenge were able to predict which new or full moon would give rise to an eclipse of the moon or the sun. Building on progressive reasoning, Thom concluded that "The 18.6-year cycle would obtrude itself . . . and an analogy with a probably earlier study of the solstices would suggest the use of a horizon mark with a back-sight."[20]

These assumptions, taken together with the speculation that sooner or later someone would notice that eclipses happened at the new or full moon nearest the date when it reached its standstills, account for Thom's later discovery of minuscule variations in the moon's movements, which he argues the astronomers marked out via the arrival of the lunar disk at particular bumps and notches on the horizon. In both Thom's and Hawkins's views, the desire to predict seasonal eclipses is the *raison d'être* for paying attention to the lunar standstills. (Cycles of eclipses that fall at the same point in the solar year would more likely have been thought to be connected with seasonal change and perhaps be easier to detect.) Both Hawkins and Thom portray the megalithic builders as progressive people who were challenged by repeatedly witnessing these impressive phenomena—people who possessed a desire to master eclipse prediction.[21]

What about the Aubrey Holes? Do they constitute evidence for eclipse prediction? Even if Stonehengers could predict eclipses, whether the fifty-six Aubrey Holes within the ditch-and-bank were used as an archaic computer to tally intervals between the dynamic encounter of the two brightest celestial luminaries remains problematic. If the Aubrey Holes were involved in such a relatively sophisticated art of eclipse predicting, then why are they positioned at the early end of the Stonehenge chronology as Table 3.1 indicates? Archaeologist Richard Atkinson suggests that rather than counting devices, the mysterious Aubrey Holes, which started out as the holes of a ring of timber posts, later became offertory pits. Archaeologists also discovered within some of them remains of burnt soil, cremated human bones, and long bone pins (probably hairpins) mixed with chalk. It is difficult to reconcile this evidence with the hypothesis that the holes ever were part of a computing device. A more reasonable idea might be that they served as symbolic doorways to the gods of the underworld. Finally, fifty-six is a very strange number, undiscovered anywhere else in the annals of Neolithic archaeology. And astronomically speaking, it turns out to make a pretty poor eclipse cycle (see Table 2.1, p. 36).

If the people of Stonehenge were watching the moon, I think they were probably more interested in charting out its phase cycle than in predicting when it would be eclipsed. Recall that the number of holes in the Y and Z circles (the concentric rings of pits located within the Aubrey Circle) total 30 and 29 respectively. These are reasonable whole-number approximations of the length of the lunar synodic month. Also, there are 59 stones, the sum of 30 plus 29, in the bluestone circle. Thirty stones make up the Sarsen Circle. That all three of these circles were set in place in the interval 2450–1500 B.C. may imply that during this period there was some interest in reckoning time by the phases of the moon.

As we cross the threshold of time into the sixth millennium of Stonehenge's existence we see that it has been interpreted and reinterpreted many times. The twentieth century—call it the age of computer-assisted quantitative science—has given us a contemporary version of Stonehenge. It emerges as a place for scientific calculating and astral predicting. Reading Hawkins and Thom in hindsight, it becomes clear that their generation—the 1950s and 1960s—was one of both controversy and embroiled discussion in the field of cosmology. At that time the steady state theory, which posited that the universe was eternal, having neither a beginning nor an ending, was crumbling beneath mounting evidence that all things came about in a violent burst of creation 10 to 15 billion years ago—the Big Bang. It was also a time when methods of making scientific calculations were undergoing drastic revision. As Hawkins remarks, there is only one way to determine whether the alignments he measured possessed any celestial significance: "We need precise measurement and comparison, a great volume of trial-and-error work—much more work than I can find time to do. We need the machine."[22] In effect Hawkins ended up using a computer to prove that Stonehenge itself was a computer.

Over the past few decades, reflection, coupled with remeasurement and analysis of Megalithic sites by interdisciplinary teams of researchers, seems to have toned down the presumed scientific motivations of our archaic ancestors.[23] The compromise Stonehenge model the new millennium seems to desire leaves room for a fusion of scientific ideology with religious worship and social concerns. Postmodern Stonehenge recognizes the diversity of the interrelated components that make up civilized culture.

Even if Hawkins and Thom overshot in their estimations of the degree of precision of ancient mensuration and motivation for acquiring celestial knowledge, both investigators should be credited with revealing a definite astronomical function for Stonehenge and other megalithic sites. The sun and moon alignments have withstood the test of time; the solar and lunar standstills probably were intended not to predict eclipses, but rather to honor the sky gods as they took up their assigned approximate places. If we insist on calling Stonehenge an observatory, then we must label it a *sacred* observatory. I am convinced it was, for some considerable period of time, a consecrated space for watching the sky, a place where cosmic encounters were celebrated because they served to call people together to conduct rites to their gods. And if the arrival of the sun and the moon in their proper places were being marked, the device that marks them becomes a timepiece. Stonehenge—unwritten and set in stone—is a calendar about as far from our own in concept, design, and appearance as we could imagine.

CHAPTER FOUR

POWER FROM THE SKY: ANCIENT MAYA ASTRONOMY AND THE CULT OF VENUS

the calendar of the Indians of New Spain ... they counted by a star ... that we call Lucifer. ... And in this land the duration of time from the day when it first appears to when after rising on high it loses itself and disappears, amounts to 260 days, which are figured and recorded. ...

—Jacinto de la Serna,
Idolatrias Supersticiones etc. del las Razas Aborigenes de Mexico, 1882

THE CIVILIZATIONS OF ANCIENT MESOAMERICA

The Western world did not become aware of the existence of an advanced civilization in the Americas until the travel-writer and diplomat John Lloyd Stephens, accompanied by artist Frederick Catherwood, toured Central America in 1839–1840. They produced two sets of volumes, *Incidents of Travel in Central America, Chiapas and Yucatan* (1841) and *Incidents of Travel in Yucatan* (1843). Coming at a time when Americans were becoming curious about what lay far west of the Appalachians, beyond the settled bounds of their relatively new nation, both books became instant best-sellers. In words and pictures Stephens demonstrated that the achievements of the ancient Maya in the fields of art, sculpture, architecture, and writing were in many ways on a par with the classical civilizations of the Old World. And he correctly attributed all these accomplishments to an indigenous race of American people when he answered his own question:

> Who were the builders of these American cities? They are not the works of people who have passed away and whose history is lost but of the same races who inhabited the country at the time of the Spanish conquest. . . .[1]

Anthropologists define Mesoamerica as the region bounded on the north by the Tropic of Cancer and on the south by northern Honduras. It was originally populated by nomadic peoples from central Asia who crossed the land bridge into Alaska more than 10,000 years ago. These early people moved with the seasons, hunting and gathering their food supply as they went; but by 2500 to 2000 B.C.—what archaeologists call the Early Formative period—isolated pockets of sedentary civilization sprouted and an agricultural system, based principally upon maize, took hold. This period also saw the beginning of pottery making and the expansion of an organized pattern of trading between villages.

It is impossible to state when the people of Mesoamerica attained that sophisticated condition of human society we call "civilization." So many factors are involved in the definition of that term, and to make matters worse, the material evidence is very scant. What Mesoamerican archaeologists call the Pre-Formative period began about 2500 B.C. (about the time the Wessex culture entered Salisbury plain) with the appearance of ceramic works. Settled village life had recently begun to develop in the "Olmec Heartland" along the tropical-forest Gulf coast of southern Mesoamerica, along with farming based on corn, beans, and squash. One of a number of Mesoamerican civilizations to develop complex art and architecture during the Formative or Pre-Classic period (beginning about 1800 B.C.), the Olmecs are well known for their distinctive art style, particularly the mysterious giant stone heads they carved, thought to be portraits of their leaders (one measures 9 feet in height!). They also worked delicate figurines out of jade—human baby faces with jaguar features like fangs and a snarling mouth. These may have represented their concepts of prehuman origins and their ideas about the relationship between human and animal forms.

By the twelfth century B.C. the great Olmec ceremonial centers of Tres Zapotes, La Venta, and San Lorenzo flourished, some of them boasting pyramids up to 100 feet tall. The highly stylized Olmec art and architecture of the Gulf Coast strongly influenced the nascent Maya civilization, which would soon grow up to the east in the Yucatán peninsula. Massive pyramids, like the one at La Venta, 25,000 square yards at the base, began to serve as the focal points of sacred ceremonial complexes. Temples and pyramids rose up in greater numbers out of the jungle.

In the Middle Formative period (1000 to 300 B.C.), settlements sprang up in the valleys of Mexico and Oaxaca in the central highlands. The civilization of Teotihuacan erected ancient Mexico's largest city shortly before the beginning of the Christian era. Here was a culture that would leave its impact on all Mesoamerican societies that followed it. Originating out of a cluster of communities in the valley of the San Juan River near the shore of highland Lake Texcoco (today 30 miles northeast of Mexico City), on a major obsidian trade route, Teotihuacan was said by the Aztecs, whose culture would develop some 1,500 years later, to be the birthplace of the gods and the place where time began. By the beginning of the Christian era it covered an area of 8 square miles and had a population of more than 100,000 people, who resided in more than 2,000 apartment compounds. The city was well planned, being divided into quadrangles by a grand 2-mile-long boulevard nearly as wide as a football field running north-south, and a shorter east-west street running almost exactly perpendicular to it. (Builders deliberately skewed the grid to fit with the sacred land- and sky-scape—see Figure 3.5). Teotihuacan's giant Pyramid of the Sun, built over a sacred artificial cave, stands 210 feet high and is 700 feet wide at the base. Great palaces housed the elite: the rulers, priests, military men, senior merchants, and administrators.

Toward the end of the Middle Formative, the first concrete astronomical achievements can be documented: the beginnings of a 365-day year, and the unique 260-day cycle, consisting of 13 numbers and 20 named days running side by side, on the first carved stelae, or upright stone slabs (see Figure 4.1). This period was also characterized by rapid advances in the arts

Figure 4.1 According to epigrapher John Justeson, this carved stela from La Mojarra on Mexico's Gulf Coast exhibits earliest form of hieroglyph for Venus (circled). Following the head of a sun-eating monster (also circled) is a Long Count date that correlates with a visible solar eclipse. The life history of the king (shown in profile wearing a huge fishtail headdress) is also given.

and sciences as well as great architecture and sculpture, all of it accompanied by increasingly complex political and social systems.

The period of greatest sophistication in the civilizations of Mesoamerica occurred between A.D. 200 and 900, a time when much of Europe slept in intellectual darkness. Called the Classic or Florescent period, this era was characterized principally by the appearance of highly organized settlements, an advanced calendar, a complex religious pantheon, and the rise of a social elite class. Nowhere in Mesoamerica were these qualities of advanced civilization more outstanding than in the land of the Maya, in the rain forest of the central Yucatán peninsula. Tikal's monumental architecture, the delicate high-relief sculpture of Copán, and the exquisite stucco work of Palenque seem unsurpassed in the New World and rival those in the Old.

An abundance of carved stelae, arranged around the plazas of the great Maya cities like so many gravestones, still greets today's tourist. But unlike Stonehenge's megaliths, these are inscribed with information (perhaps much of it political propaganda) certifying the exploits and descent from the gods of great kings and queens. Their effigies are portrayed alongside complex calendrical dates that tell the time of year and position in the sacred 260-day round of each significant event—birth, accession to the throne, captures of enemy warriors, betrothal, and so on. But these zealous pursuers of time wrote even more on their standing stones: the correct phase of the moon, its position in the zodiac, even the count of days since the time of creation (a number running into the millions!)—all corresponding precisely to each happening.

The inscriptions also speak of the gods, their ancestors, but it is difficult to know whether the many gods of the Maya—among them the god of rain and fertility, the fire god, the wind god, and the corn god—were specific anthropomorphized supernaturals who represented individual aspects of nature, or mana-like impersonal spiritual forces that made up an animate material world. Perhaps they were intended as performers who personify or masquerade as the forces of nature.[2]

A combination of circumstances, among them environmental mismanagement, perhaps accompanied by a popular revolution, and possibly a change of climate, led to the precipitous decline of Maya culture in central Yucatán. By the tenth century, activities shifted north to the great cities of Chichén Itzá and Uxmal. Why the fall was so widespread and complete still remains a mystery. But we can be sure that the Maya to a large degree gave up their obsession for carving calendar dates on stelae, lavishly entombing their dead kings, and continually refurbishing their massive architectural works—all habits of the Classic period. Basically they fell back to a simpler existence. By the time the Spanish invaders arrived on the scene early in the sixteenth century, native villagers found themselves in relative disunity, a factor that may have complicated and prolonged the conquest.

A few centuries after the collapse of the Maya, the great Aztec state rose up in the highlands to the west and dominated Mesoamerican cultures from about the middle of the fourteenth through the beginning of the sixteenth centuries, when it was abruptly confronted by the Spanish invasion. When he entered Tenochtitlan (today's Mexico City), Cortés, the celebrated Spanish conquistador, who was hailed as the prophesied returning feathered serpent-god Quetzalcóatl, broke the bond of the ruling Aztec hierarchy and gained the day's advantage in relatively short order. What little remained of one of the great civilizations of the world had become almost totally muted in less than a single generation.

We don't know why the Maya, more than most ancient civilizations, got so carried away with looking at the stars, but by the end of the Classic period they had managed to develop a recyclable Venus calendar that was accurate to one day in 500 years, and an eclipse warning table that still functions today. They devised their own zodiac as well as a table to follow Mars (which suggests they may have been as fascinated with retrograde motion as Kepler). To accomplish these impressive intellectual feats, they created a sophisticated system of mathematics utilizing place value and the concept of zero, which greatly facilitated their computations. They projected their astronomical tables thousands of years forward and backward, even to eras preceding the creation of the world. How modern scholars have come to know these facts is as fascinating as the facts themselves.

To comprehend the nature of Mesoamerican astronomy in general, in many cases we need to rely on statements written down by historians of the post-Conquest or Colonial period, many of whom, regrettably, had not recorded their observations very carefully. Bent on religious conversion, these holy men were concerned with native customs only to the degree that they might be employed in the acculturation process. For example, the Spanish historian Torquemada, writing a century after the conquest, tantalizes us with this statement about the astronomical pursuits of Netzahualpilli, the king of Texcoco:

> They say he was a great astrologer and prided himself much on his knowledge of the motions of the celestial bodies; and being attached to this study, that he caused inquiries to be made throughout the entire extent of his dominions, for all such persons as were at all conversant with it, whom he brought to his court, and imparted to them whatever he knew, and ascending by night on the terraced roof of his palace, he thence considered

the stars, and disputed with them on all different questions connected with them.[3]

Exactly what was this astrologer-king observing? What equipment was he
using? How did he record his knowledge? These are all questions twenty-
first-century students of astronomy are bound to ask, and sadly, none of
them can be readily answered.

One text even goes so far as to mention specific native constellations. In
the *Crónica Mexicana*, the post-Conquest historian Alvarado Tezozomoc
gives an account of the formalities that took place upon the election of
Montezuma Xocoyotzin, king of the Aztecs. Following a long list of religious duties, he was exhorted:

> especially to make it his duty to rise at midnight and to look at
> the stars: at *yohualitqui mamalhuaztli*, as they call the keys of St.
> Peter among the stars in the firmament, at the *citlaltlachtli*, the
> north and its wheel, at the *tianquiztli*, the Pleiades, and the *colotl
> ixayac*, the constellation of the Scorpion, which mark the four
> cardinal points in the sky. Toward morning he must also care-
> fully observe the constellation *xonecuilli*, the "cross of St. Jacob,"
> which appears in the southern sky in the direction of India and
> China; and he must carefully observe the morning star, which
> appears at dawn and is called *tlahuizcalpan teuctli* [see Figure 4.2].[4]

But the words of historians passed down to us about Maya sky watching
are even more far-reaching than this, for they tell of a people who used
mathematics and writing to explore the science of the sky, which seems to
have been based on the precise timing of celestial events. Of the Maya
astronomers, Bishop Diego de Landa, on the scene in Yucatán within a
decade of the conquest, says:

> They have their year as perfect as ours.... They used as a guide
> by night, so as to know the hour of the morning star, the
> Pleiades and the Gemini...[and they made use of] certain
> characters or letters which they wrote in their books their
> ancient matters and their sciences....[5]

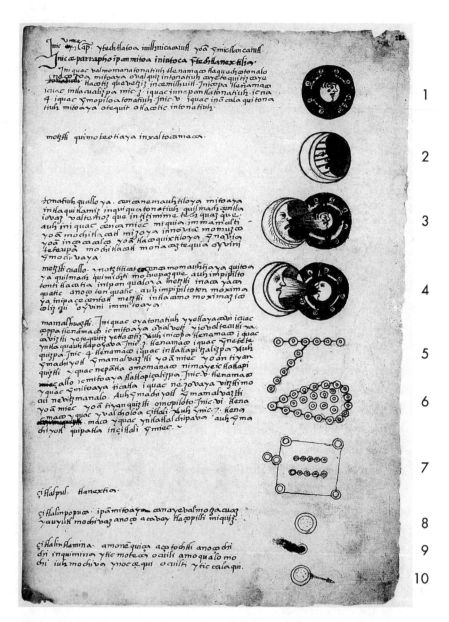

Figure 4.2 Sky objects in the *Primeros Memoriales*, an early Spanish document from post-Conquest Mexico. (1) *Tonatiuh* (sun), (2) *Meztli* (moon), (3) solar eclipse, (4) lunar eclipse, (5) *Mamalhuaztli*, the Fire Drill (Orion's belt and sword?), (6) *Tianquiztli*, or the marketplace (the Pleiades), (7) *Citlaltlachtli*, or the ballcourt (Gemini?), (8) *Citlalpol* (Venus), (9) *Citlalpopoca*, or smoking star (comet), (10) *Citlaltlamina* (shooting star), (11) *Xonecuilli* (Little Dipper?), (12) *Citlalcolotl* (Scorpio?). Other sky phenomena depicted include snow (rare in Mexico City), lightning, and a rainbow.

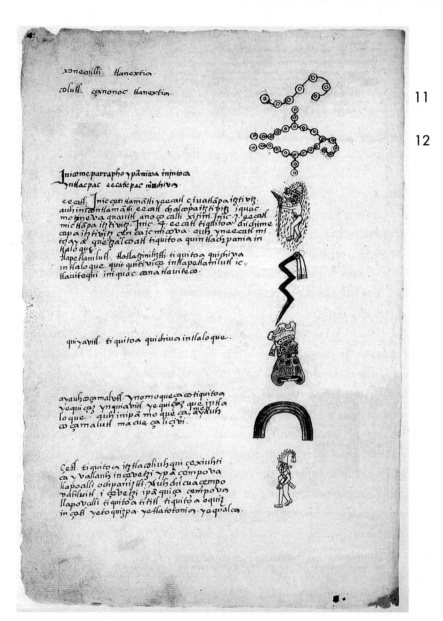

xoneoilli. tlanextia.

coluhl. cononoc tlanextia.

11

12

Iniome parrapho ypamitia injnfoa
yntlacpac eecatiepac ninchiva.

eecatl. Jnic can tlamatli yeecatl ciuatlapa ightivh.
auh inicantlamatli eecatl chalcopaiztivit. iquac
motineva qnauitl anoco calli xitini. Jnic 3. eecatl
mictlapa ihtivih. Jnic. 4. eecatl tiquitoa diichime
copa ihtivih cenca icmicova. auh yneecatl mi
tiday a quetlalcoatl tiquitoa quintlachpania in
tlaloq.
tlapetlaniluh. tlaxtlaximiliztli tiquitoa quichiva
intlaloque. quit quitivica intlapetlaniluh ic
tlauiteqiii iniquac cana tlauiteco.

quiyavitl. tiquitoa quichiua intlaloque.

ayauhcocamaluh. Ynomoquecacotiquitoa
yequiicaz ynqniiavitl yequicoz que intla
loque. auh inipa mo que ca, ayauh
cocamaluh maciacaliciivi.

Cetl. tiquitoa iztlacoliuhqui cexiuhti
ca yvallauh incovetzi ypacompova
tlapoalli ochpaniiztli. Auh chicuacempo
valiuiut i covetzi ipaquica cempova
tlapovalli tiquitoa tititl tiquitoa oquin
incetl yetoquizpa yetlatotonioa yaqialcos.

Figure 4.2 (Continued).

The chronicler Jose de Acosta gives details about these texts, fashioned of lime-coated bark, describing them as:

> books of leaves, bound or folded after a fashion, in which the learned Indians kept the distribution of their times and the knowledge of plants, animals, and other things of nature and the ancient customs, in a way of great neatness and carefulness.[6]

In the central Mexican books called codices (a misnomer for folding screen documents made of deerskin), a number of pictures help to illuminate our understanding of the techniques and objectives of practical astronomy in Mesoamerica. For example, the Codex Mendocino, or Mendoza (1831), a picture book produced shortly after the conquest, tells about various aspects of the lives of certain members of the Aztec noble class in their capital city of Tenochtitlan. The first three parts of Figure 4.3 are taken from it. Figure 4.3a contains adjoining captions written in Spanish and based on interviews with native informants. The seated priest is said to be "watching the stars at night in order to know the hour, this being his official duty." An inverted hemisphere studded with stars, symbolized by half-shut eyes, hangs over his head. In Figure 4.3b, which appears adjacent to it, another priest is beating on the *teponaztli*, a wooden drum, to announce the time of night as determined from the observations of the first priest. Finally, Figure 4.3c informs the reader that the time of night is recognized as suitable for the performance of an obvious agricultural function. These drawings emphasize the utilitarian role of nighttime sky watching among urban highland people, a theme we will follow to a greater extent among the Maya of Yucatán.

One of the great mysteries of New World astronomy lies in comprehending just how these people could have charted their tropical sky. Theirs was a low-tech culture that possessed no calculating devices, telescopes, or measuring apparati. But some pictorials are more explicit; thus, in Figure 4.3d, from the Bodleian Codex, we find a picture that has been well established by epigraphers as a logo or place name representing the ancient town of Tilantongo, near modern Oaxaca. But the picture also may contain some insights into astronomy. We see the profile of a man situated in a chamber within a temple. He peers out the doorway and looks over a pair of crossed sticks, as if to mark the place of an astronomical event on the horizon. The

Figure 4.3 Chronology represented one of the earliest demands made upon New World astronomers. Views (a), (b), and (c) from the Mendoza Codex illustrate the role of nighttime observations in timekeeping. (d) An astronomer looking over a pair of crossed sticks perched in the doorway of a temple (Bodleian Codex). (e) A smoking star, possibly a meteorite fallen from the star-studded heavens, is connected with a battle. (f) An eclipse of the sun, which was actually visible in the year indicated in Tenochtitlan (Codex Telleriano Remensis).

outside of his temple is studded with the same star symbols we saw in Figures 4.3a, b, and c, which suggests that the building in which he is situated also may have functioned as an astronomical observatory. Is Tilantongo the place where the house of the skywatcher is located? Are the sticks actually a rudimentary measuring device? What horizon event is being witnessed? Does the star temple have a special orientation toward the object on the horizon?

Using a pair of notched sticks, an observer can determine the position of an object near the horizon with great accuracy. As depicted in Figure 2.6b (see page 22) a pair of vertical markers could have been set in fixed locations, one as a backsight and the other as a foresight, to record the position of an astronomical body. When that body returned to this position between the notches, the astronomer could determine the length of its cycle. Perhaps a prominent feature in the landscape functioned as a natural foresight. In either case, any observatory edifice would have to be preferentially oriented so that it faced that part of the landscape where the event occurred. As archaeologists have yet to excavate a set of cross-sticks, such evidence is circumstantial, yet I believe this sort of alignment scheme is plausible. And, as we shall see, there are plenty of disoriented, astronomically aligned buildings in Mesoamerica. A few other astronomical references depicted in the codices are pictured in Figures 4.3e and f.

Clearly what is singular about the Maya, above all other Mesoamerican cultures, is that they possessed a writing system. Diego de Landa tells the sad tale of the fate of the written legacy of the ancient Maya. He refers to a great book burning that took place at the Yucatán city of Maní in 1555:

> We found a large number of books in these characters [hieroglyphs] and, as they contained nothing in which there were not to be seen superstition and lies of the devil, we burned them all, which they regretted to an amazing degree, and which caused them much affliction.[7]

One look at the frightening effigy of Venus-*Tlahuizcalpantecuhtli* (Figure 4.4) (from the central highlands or the equally grotesque Kukulcan of the Maya pictured in Figure 4.8 on page 116) gives us pause to realize that in

Figure 4.4 *Tlahuizcalpantecuhtli,* the Venus deity of central Mexico, holds up the sky (from the Codex Borgia).

his own eyes Landa, bent on converting the savages of his newfound land, was perfectly justified in destroying the records and pictures of those who would worship a god so alien from his own.

As a result of the zeal of the conquerors, fragments of but four original Maya codices—the Dresden, Paris (Peresianus), Madrid (Tro-Cortesianus), and Grolier—survive today. (The Madrid is seven yards long!) Most are named after the places where they now reside. The two dozen other codices that survive from central Mexico make up the entire corpus.

The Maya books, the only ones that contain hieroglyphs, have the most to say about sky watching; therefore, once we get a feeling for how the Maya kept time we will find a wealth of detail that indicates just how deeply Maya astronomers penetrated their tropical skies.

MESOAMERICAN TIMEKEEPING

We have already learned in Chapter 1 that chronology—setting up an ordered system of timekeeping to administer to the needs of the state—was one of the basic motives for ancient sky watching. This was especially true for the Maya. Their written works strongly imply that they possessed a deep desire to create an elaborate and precise calendar, a desire that developed into an obsession rarely paralleled in human history.

As in ancient Mesopotamia, Maya writing began as pictures and gestures that later were altered into signs that took on more abstract forms. Take the signifier for moon (displayed in Figure 4.5), whose phase and orientation highland Guatemalan farmers still indicate by hand gestures. In written form the moon sign evolved into an ever more stylized crescent form. One of the great breakthroughs in Maya epigraphy (the study of writing) occurred in the mid-1970s when a team of interdisciplinary scholars, building on earlier research by Maya linguists, demonstrated that a significant portion of classic Maya script was phonetic; that is, the written symbols stood for sounds or particles of speech that made up whole words. In other words Maya writing was like ours, with homonyms, synonyms, and a grammar all its own. It differed only in that it had a syllabary (with over 1,000 entries) rather than an alphabet to hold it all together. Recounting the history of this decipherment would be too lengthy a diversion to embark upon here, but it is safe to say that it all started a century ago when scholars first recognized, among the orderly columns of characters written in the codices and carved on the monuments, repeating patterns of certain glyphs and numbers—the Maya cycles of time.[8]

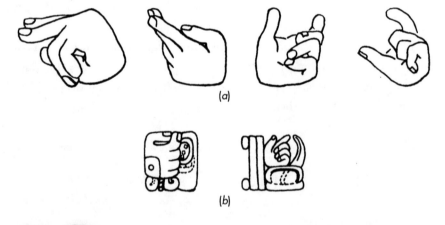

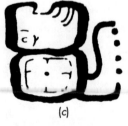

Figure 4.5 From hand gestures to writing. (a) Contemporary hand signs relating to the moon: moon going down; rising young moon; moon retaining water, or dry season; moon letting water escape, or rainy season. (b) Maya hieroglyphs: the moon at new phase; the moon at gibbous phase, 17 days after new moon. (c) A more complex hieroglyph, meaning West, or "where the day is completed," a closed fist over u kin, or day, sign.

For the Maya a single word, *kin*, signified time, day, and the sun. Its meaning and basic glyphic form (Figure 4.5c) suggest that the art of time-keeping was intimately connected with the practice of astronomy. The directions of the petals of the floral design that comprise the day glyph may correspond to the standstill positions of the sun at the horizon.

Where would our astronomy—indeed, most of our sciences—be without quantitative forms of expression? Facility with mathematics is paramount in any culture that would develop a complex astronomical chronology, and it is in this area that the Maya seem to have outshone all their worldly contemporaries. The introduction of the concept of zero into any mathematical system lends considerable facility to the performance of simple mathematical operations, especially when the position of a number in a series determines its value and contribution to the whole. (Anyone who still doubts the computational advantage of a mathematical system with zero in it might try adding or subtracting in Roman numerals!) The Maya expressed large numbers in a notational system utilizing place value and a zero, which was quite like our own system that developed in the Middle East after the fall of the Roman Empire. For example, when we write the number 365 usually we are not very conscious of its meaning. This compact notation tells us that the number is the sum of 5 ones, 6 tens, and 3 ten-times-tens. We use a decimal or base-10 place system, undoubtedly because our ancestors counted articles on the fingers of both hands; thus we define a completed unit as that which is equivalent in content to the number of fingers possessed by a whole normal human being. Adding and subtracting become simple operations when we can borrow or lend units readily among orders.

While place value and the concept of zero are two advanced features that the Maya system shares with our modern Arabic notation, there are two primary differences. First, the Maya numbers were written vertically rather than horizontally, with place value increasing from bottom to top, and second, the base of the system was 20 instead of 10. Can you guess why?

In base 20, or vigesimal notation, every higher order in the system is cast in powers of 20 rather than 10. For example, in Maya the number shown in the fifth vertical box in Figure 4.6 (which we will write as 8.2.0) would stand for zero ones plus two twenties plus eight twenty-times-twenties. Expressed in our decimally based system, this adds up to 3,240. Note that the zero lies in the lowest numerical order at the bottom in this particular

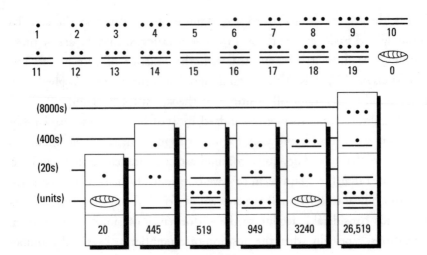

Figure 4.6 The Maya numeration system: Dots represent ones, the bars are fives, while the closed fist, or stylized shell, stands for zero. At the top of the chart are the numbers from one to nineteen and zero. Below, several examples of larger numbers are given. These numbers are tallied vertically in base 20. The position at the bottom is the unit position, above it is the 20 position, then the 400, with the 8,000 position on top. The third column from the left, for instance, is figured as 1 × 400 + 5 × 20 + 19 = 519. When the Maya counted days, however, the 400s became 360s and the 8,000s became 7,200s, etc. Our number 519 in this case becomes 479. (For practice, try translating the other numbers on this page from trade count to time count.)

notation. It is represented by a closed fist or stylized conch shell (which that shape resembles), the fist denoting completion.

An essential difference in Maya notation is that when used as a time count—as opposed to a trade count in which articles or things are being tallied—the third place in the sequence of dots and bars becomes 360, or 18 × 20, rather than 400, or 20 × 20; and the next higher unit becomes 20 × 360 or 7,200, and so on. This would have made the counting system more user-friendly when tallying units such as seasonal years. (The Maya employed a count for the year, consisting of 18 months of 20-day duration plus an added month of five days.) Thus, if 8.2.0 were intended to represent elapsed *days* rather than counted coconuts, it would translate, in our time currency, to 2,920 days—or roughly eight years, two 20-day months, and zero days. If this seems odd, look at the special way *we* deal with time: Our Babylonian-derived sexagesimal system accords 60 seconds to the minute and 60 minutes to the hour, rather than one hundred.

Maya chronology, which makes extensive use of these numbers, is based on cycle building; that is, on the accumulation of smaller cycles to make bigger ones. Not all the cycles in the chain were purely astronomical. The most important time cycle, the 260-day sacred round (called the *tzolkin*, or count of the days) linked together two smaller cycles of 13 numerals and 20 day names (not unlike the way our 7-day week and 30 numbered days of the month go together, except they made no exceptions like February). If all our months had 30 days and we matched each consecutive day number with the 7 consecutive named days of the week, then a complete cycle would run its course in 7 × 30 or 210 days before a given day would be labeled with the same number. This is exactly how the *tzolkin* works.

No other civilization in the world used a 260-day cycle, and we are not even sure how this peculiar period rose to prominence in Mesoamerica.[9] Nonetheless, this vital segment of the Maya calendar remains a living relic. Used largely for divinatory purposes, it is still employed among some remote people of the highland region of Guatemala. And lucky and unlucky day names still function today as they likely did in ancient times.

In Momostenango, for example, a diviner answers questions about illness, land disputes, lost property, dreams, omens, and so forth posed to him by a client. He draws a handful of seeds and crystals out of his bag and spreads them on his table, arranging them in rows consisting of small piles. Then he addresses them, asking for light and clarity as he counts out the days by name and number: 1 *Quej*, 2 *K'anil*, 3 *Toj*, 4 *T'zi*...all analogs of ancient Maya day names in the *tzolkin*. "Will this woman be given in marriage?" The pile of crystals and seeds at which he stops gives the answer, speaking to him through his blood, he says.[10]

If modern Maya divinations are anything like those of the Classic era (and there is little reason to think they are very different), then we are drawn to the conclusion that the keeping of the ritual books, filled with their multitude of cycles, was an act more closely paralleling astrology than scientific astronomy. Here was a scientific pursuit not unlike our own, except that it was driven by different ends and different needs. Theirs was a religious agenda, quite distinct from modern scientists' professed goal of understanding nature for its own sake. In my view this only deepens the mystery of why the Maya watched the sky and kept time so precisely.

ASTRONOMY IN THE MAYA CODICES

Compared to the culture that erected Stonehenge, the Maya have left us a written legacy—an advantage we did not possess when we were struggling to understand the status of astronomy in Bronze Age Britain. The crowning achievement of Maya astronomy resides in the Dresden Codex, a Maya picture book produced in northern Yucatán about the eleventh century and unearthed in a German library eight centuries later. Devoted primarily to prognostications and divination, it is believed to be an updated copy of another folded-screen document produced a few centuries earlier. The Dresden, as we learned, is one of just four surviving fragments of Maya books, all of them having been delivered by conquistadors to the enlightened European nobility as representative New World curiosities. The Dresden is in the best shape. Its glyphs are painted with a fine brush in vivid reds and blacks, while other figures are rendered in yellow and blue. Looking at the Dresden is a strange experience for us. We are puzzled by the juxtaposition of huge numbers and phonetic writing with grotesque pictures of vile, half-human, half-animal creatures. But a penetrating look at these exotic painted panels rewards us with a banquet of food for astronomical thought.

An eminent Mayanist once called the Venus Table on pages 24 and 46 to 50 of the Dresden Codex "a subtle and mechanically beautiful product of Maya mentality."[11] However, he went on to say that the general layout of these pages proves that the Maya were not so much concerned with glorifying their own intellectual achievements as they were with venerating their gods. If there is one lesson to be learned from the studies of the astronomies of ancient cultures—the lesson of Stonehenge—it is that other people do not necessarily see the world the way we do. Thus in the Dresden Codex Venus Table we discover numbers that seem to have something to do with Venus, except that they do not always look like the correct numbers; that is, they are not necessarily the numbers we might use to follow Venus's cycle of appearances and disappearances. The same can be said of the arrangement of numbers in the so-called Eclipse Table on pages 51 to 58 of the Dresden. A brief trek through the Eclipse Table, which is easier to understand, will serve as a warm-up before we tackle the exquisite detail in the Venus Table, which ties in with one of the principal celestial concerns of the Maya—the worship of the planet Venus.

Figure 4.7 shows a segment of the Eclipse Table. You read it by scanning across the top half of each consecutive page; then you return to the first page of the table and continue across the bottom half. The dot-and-bar and hieroglyphic text is punctuated by pictures that suggest eclipses are involved here, for example, a dragonlike creature eating the symbol for the sun (not shown), and curious moon and sun glyphs presented in half light/half dark backgrounds. But, as we shall see, it is the numbers lined up across the bottom of each half page that clinch the argument—numbers familiar to anyone who watches the phases of the moon with the intention of predicting when it will be eclipsed. They translate as a repeated chain of 177s followed by a 148, then a picture.

A few lines above each of these numbers we recognize another series of dot-bar numerals, written in time count. For example, if you look carefully just above and in between the bottom pictures in Figure 4.7 marked as pages 52 and 53 you will find the numbers 6,408, 6,585, 6,762, 6,939, 7,116, and 7,264. It is easy to see that if we add the lower number of a given column to the upper number of the previous column we arrive at the upper number in the next column. For example, 6,408 + 177 = 6,585, and 7,116 + 148 = 7,264; therefore, the upper numbers must be the totals or cumulatives one arrives at by repeated addition of the lower numbers. The whole scheme is laid out in Table 4.1.

To the attentive reader, two observations immediately suggest that these computations are related to eclipses: first, the appearance in the table of the *saros* interval of 6,585 days (see p. 36 and Table 2.1—this was the famous Chaldean 223-month, 18-year seasonal eclipse cycle), and second, the near equivalence of 6 lunar synodic months to 177 days (actually 6 lunations = 177.18 days) and 5 lunations to 148 days (5 moons = 147.65 days). The number 177 is a very likely one to grace an eclipse table, for it is also close to exactly 6½ draconic months.

The intervals between the nine pictures are 1,742, 1,034, 1,210, 1,742, 1,034, 1,210, 1,565, and 1,211 days, all of which can be recognized as real eclipse cycles from our discussion of unaided-eye astronomy (again see Table 2.1). Finally, the total number of days in the table is 11,958 (about 33 years) or very nearly 405 lunar phase cycles (405 lunar synodic months

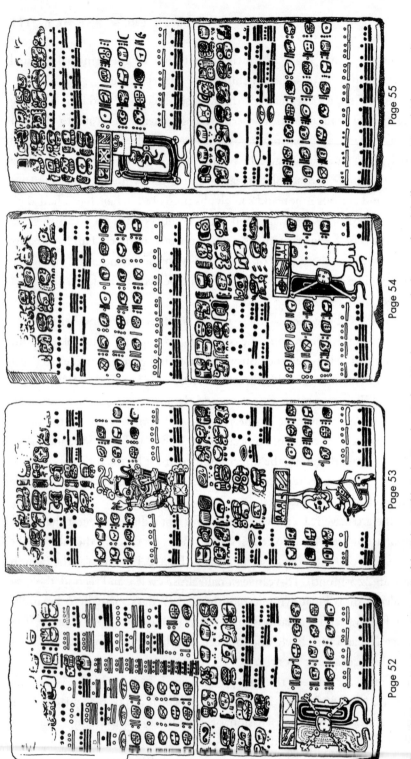

Page 55

Page 54

Page 53

Page 52

Figure 4.7 A section (pages 52 to 55) of the Eclipse Table in the Dresden Codex. Numbers across the bottom of each half page are 5 and 6 lunar synodic months and the pictures represent (usually) ominous eclipse events.

TABLE 4.1. INTERVALS AND CUMULATIVES IN THE ECLIPSE TABLE OF THE DRESDEN CODEX

177	177	178	6,585
177	354	177	6,762
148	502	177	6,762
Picture		177	6,939
177	679	177	7,116
177	856	148	7,264
177	1,033	Picture	
178	1,211	177	7,441
177	1,388	177	7,618
177	1,565	177	7,795
177	1,742	177	7,972
177	1,919	177	8,149
177	2,096	277	8,326
148	2,244	148	8,474
Picture		Picture	
178	2,422	177	8,651
177	2,599	177	8,828
177	2,776	178	9,006
177	2,953	177	9,183
177	3,130	177	9,360
148	3,278	177	9,537
Picture		177	9,714
177	3,455	177	9,891
177	3,632	148	10,039
177	3,809	Picture	
177 (178)	3,986	177	10,216
177	4,163	178	10,394
177	4,340	177	10,571
148	4,488	177	10,748
Picture		177	10,925
177	4,665	177	11,102
177	4,842	148	11,250
178	5,020	Picture	
177	5,197	177	11,427
177	5,551	177	11,781
177	5,728	177	11,958
177	5,905		
177	6,082		
148	6,230		
Picture			

= 11,959.89 days). Exemplifying the Maya passion for recognizing inter-locking time cycles, this number also meshes perfectly with the 260-day cycle or *tzolkin* (46 × 260 = 11,960 days = 405 moons) used by Maya day-keepers in their prognostications; that is, it can be used to recover the same day of the *tzolkin* with only a slight change in the phase of the moon. These conclusions lead us to a simple hypothesis: Pages 51 to 58 of the Dresden Codex represent an Eclipse Table consisting of groups of five and six moons, the eclipses occurring at the positions of the pictures in the table. But since ritual calendar dates of the *tzolkin* constitute the bulk of the table, the document must have been intended to record the dates of actual eclipses in the ritual calendar for divinatory purposes. (You can make your own eclipse table. See Appendix A.)

Is Dresden 51 to 58 a record of eclipses already witnessed or is this sec-tion of the document intended to warn of possible future eclipses? And, if so, what kind of eclipses? There seems little doubt that the Maya sought to predict eclipses because of the disaster that they believed threatened them on such occasions—omens in the table read "woe to life," "woe to earth," "woe to the seed." The pictures in the table look ominous enough, at least to us. Of course, predictions must be based upon recorded observations of actual eclipses that occurred in Yucatán when the Maya priests did their work. Though scholars are not in agreement over which particular set of eclipses (lunar or solar) was being observed,[12] the data we have given ought to provide enough evidence to support the hypothesis that the Dresden Eclipse Table was devised for warning of the possible occurrence of such phenomena.

What kind of eclipses? My own opinion is that both lunar and solar observational eclipse data can be utilized to construct a semblance of the Dresden Table. The hypothesis that lunar data were actually used seems much simpler. In the course of the thirty-three years spanned by the table, the number of such eclipses observable in Yucatán would have been signif-icant enough to enable a single priest to draw up the table. If we assume that solar eclipses alone were used—and this certainly cannot be ruled out—we must extrapolate the base of observations backward many centuries in order to derive the relevant intervals.

The Venus Table, which immediately precedes the Eclipse Table in the Dresden Codex, exhibits a similar format—pictures and intervals. For the

Maya the importance of Venus, above all planets, cannot be overstated. They variously called it *noh ek* (great star), *chac ek* (red star), *sastal ek* (bright star), and *xux ek* (wasp star). In central Mexico, Friar Toribio Motolinía tells us that "next to the sun they adored and made more sacrifices to this star than to any other celestial and terrestrial creature" and "they know on what day it would appear again in the east after it had lost itself or disappeared in the west . . . they counted the days by this star and yielded reverence and offered sacrifices to it."[13] Why all this attention? Recall how our own earth-based observations reveal that planet's uniqueness. Besides Mercury, it is the only bright planet—brightest in the sky—that appears closely attached to and obviously influenced physically by the sun. Venus announces sunrise in the morning and rises from the ashes of the deceased solar luminary as darkness approaches. Venus, as Kukulcan (Quetzalcoatl in central Mexico), the feathered serpent deity, is aptly named, for he weaves together aspects of creatures who fly in the sky and crawl upon and beneath the earth.

A searching look at the Venus Table in Figure 4.8 is enough to put off any serious student of modern science. Why does Kukulcan, the Venus god, seem so demonic in appearance as he dons different outfits across the middle band of pictures? Who are those hideous half-human creatures across the top shaking their lidded jars? And why all those speared dead animals and humanoids sprawled across the bottom panels? Above all, what place does bizarre imagery like this have in an exalted table that purports to predict sky events with unerring precision? We see no celestial metaphors (if indeed that is what is being represented here) on our modern weather maps or medical diagnostic charts. Once again our inherent naïveté about how others see and interpret the world can spoil an appreciation of what their science was about.

It is only when we descend into the intricate detail of each of the five pages of the Dresden Venus Table that we begin to capture the mind of the astronomer at work. Only then do we penetrate the Maya intellect and cut to the core of their worldview. After searching through its pages, no reader will doubt that the wise scholars who designed the Venus Table were conducting a dialog with the sky both in the poetic meter of myth, as our Western ancestors once did, and in the precise and rigorous language of mathematics, the way modern astronomers still express what they see hap-

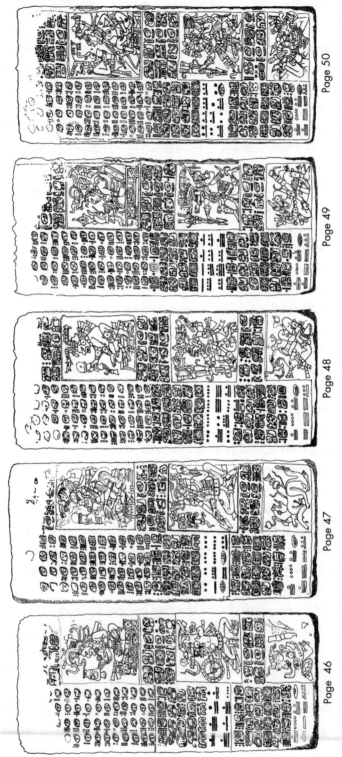

Page 50   Page 49   Page 48   Page 47   Page 46

Figure 4.8 Venus in a Maya sacred book. The five-page Venus Table on pages 46–50 of the Dresden Codex showing pictures of the Venus god flinging arrows and his events, dates, intervals, directions, and resulting omens.

pening in the sky. What astonishes us is the way the Maya sages astutely mixed magic with mathematics.

To decode the Venus Table we need to recognize two of the characteristics of the rhythm of Venus's movement that seem to have captivated the Maya: first, the division of its 584-day cycle into *four* parts, and second, the fact that each aspect of Venus repeats itself in *five* of those 584-day periods. The latter was indelibly impressed on the Maya cyclic mentality because five Venus periods happen to coincide almost exactly with eight seasonal years. This means that when, after eight years, all Venus phenomena repeat themselves, they do so on the same date of the year.

As we can see in Figure 4.8, each of the five pages of the table consists of a set of three pictures arranged one on top of the other, cemented together by hieroglyphic symbols. To the left of each lie long columns of numbers and calendrical glyphs. At the bottom of each page are four two-digit numbers. The leftmost of these numbers, 11.16, equals 236. Read in similar fashion, the second number is 90, the third is 250, and the last entry is 8.[14] Each of the five four-set blocks adds up to 584, a strong hint that the table deals with Venus. Recall that this is the time it takes the planet to pass from morning star through evening star and back to morning star again. The quadripartite division of the Venus synodic period emphasizes the nature of the Venus cycle as the Maya saw it, comprised of (1) the interval of appearance as morning star (which they put at 236 days), (2) a lengthy period of disappearance in the glare of the sun (they assigned it 90 days), (3) the appearance interval as evening star (250 days), and (4) a second, relatively brief disappearance (8 days).

As in the Eclipse Table, long-term time flows from left to right across the five pages of the table, with the pictures always following the eight-day disappearance of Venus before his predawn return to the sky. These middle pictures evidently represent the dazzling appearance of Venus as morning star. They depict five Kukulcans, the feathered serpent deity who represents Venus, all garbed in different outfits, one for each of the *five* manifestations of the full Venus cycle. (Remember the five shapes of the movement of Venus that happen over the course of eight years we plotted out in Figure 2.14?) The half man–half jaguar Venus on page 48 has been identified with an Aztec deity named Yoaltecuhtli, or guardian of the night sun, an apt role for Venus's alter ego. Garbed in five different costumes, Kukulcan flings his omen-bearing spears, bright darts of light that attend his predawn return to the world, at five different victims (they may represent other gods of nature) who lie impaled below him in each panel. For example, on page 49 squats an amphibian-headed creature; he may be a tortoise, a bringer of rain. A contemporary Maya legend has it that when the tortoise god weeps for the plight of the farmers, his tears will cause the rain to fall. To reciprocate, the serious Maya nature worshipper is still required to warn all turtles to find safety at the time the remnants of last year's cornfield are burned off in anticipation of the rainy season. Also on that same page, the central spear thrower is a blindfolded white god wearing a jaguar's skin. In this guise Venus may be the god of cold weather and another form of god of the dawn. (Like many of the gods of the classical world, Mesoamerican gods often took on multiple roles.) Finally, the five pictures across the top probably represent offerings made to the Venus deity. Except for the upper frame on the last Venus page, which depicts two figures who seem to be discussing the situation, these panels show seated figures offering up incense, plants, and so on.

There is more quintessential rhythm-making going on in the Dresden scheme for naming the actual days when the Venus events take place. The top section of each page, the first thirteen horizontal lines (the top one or two are effaced), consists of an array of hieroglyphic representations of the twenty named days of the Maya "week," each preceded by one of the thirteen numbers—the 260-day *tzolkin* with which we are already familiar. As I suggested earlier, we can think of these entries in the same way we might think of labeling an astronomic event by the day of the week and month it occurred; for instance, "a full moon happened on Friday the thirteenth." The arrangement of these named days and numbers, each one of which is aligned with a time interval at the bottom of its vertical column, indicates that the user of the table would have entered Venus time on the first horizontal line at the top of the first page, his or her eye passing all the way across each page to the end of the first line of the fifth page. Then the user would proceed to the second line of the first page, and so on until the thirteenth line on the fifth page is reached. The completed table, then, actually comprises 13 lines × 5 pages × 584 days per page = 37,960 days, or about 104 years. (We call this period a Great Cycle because it is a whole number of 584-day Venus, 260-day ritual, and 365-day year cycles.) At the end of a Great Cycle, the table can be reentered at its beginning without losing a step, provided suitable minor corrections are made. As in the Eclipse Table, a page preceding the Venus Table tells how to assess these rather complex corrections.

What do the other symbols in the Venus Table mean? The most important ones that lie between the thirteen horizontal lines of day names and numbers and the interval numbers include a set of glyphs that indicate the direction in which Venus is moving during a given interval of its orbit, and another set that tells where it will be seen. Threaded across the middle of each page are forms of hieroglyphic symbols of the planet Venus; they are labeled in Figure 4.9. Some look like a pair of eyes gazing out at us from the complex of numbers, glyphs, and pictures in the table. Notice that this same glyph is suspended from the headdress of the Venus deity on page 47 immediately to the right of the Venus symbols tabulated on that page.

What do Kukulcan's flinging spears bring in his wake? As we will see later, we find the Venus glyph elsewhere in Maya writing, sculpture, even painting, and it is in these contexts that we discover further clues about its meaning. Because it will help us understand what the Maya observed, I have made a literal translation of a short segment of the table (Figure 4.9) to accompany Figure 4.8. It begins with the set of four day names in the 260-day round in the tenth line of the first page. For simplicity, I have omitted much of the extraneous information, such as specific colors the Maya associated with the directions, month names, and so on. For each day name (one to twenty) and associated day number (one to thirteen), entries in the table announce key positions in the Venus cycle; they tell which way Venus was moving, in what direction it appeared or disappeared, and how long it was absent. It reads: *And then on the day named 10 Cib, moving to the north* (six lines below and in the same vertical column), *Venus* (two lines farther below in the same column) *disappears in the east* (seven lines farther below in the same column) *having been seen for 236 days* (two lines farther below). The next stage of the cycle happens on 9 Cimi (line 10, second vertical column), and the text continues in the same manner: *And then on 9 Cimi, in the west, Venus reappears, from the north, having been absent 90 days. And the next: And then on 12 Cib* (line 10, third column) *moving to the south, Venus disappears in the west, having been seen 250 days. And then on 7 Kan, in the east, Venus reappears from the south, having been absent 8 days.*

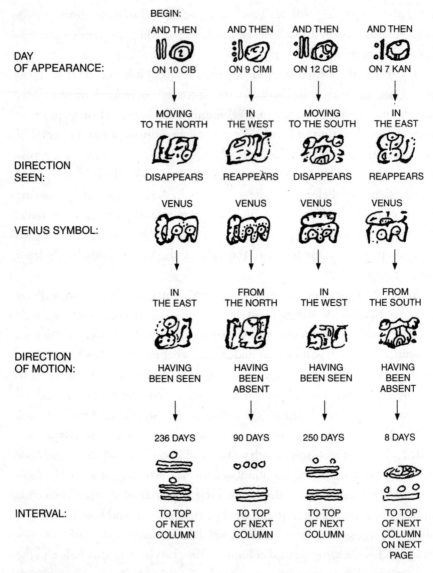

Figure 4.9 Translation of a portion of the first page (page 46) of the Venus Table, showing the divisions of time over a single 584-day cycle. The whole table incorporates sixty-five such cycles before it repeats itself.

It is at this crucial stage of the cycle that bright Venus reappears in the sky as morning star. The heliacal rise event is represented in the table as the image of the shield-bearing Venus god in one of his five manifestations throwing darts of light. A block of a dozen glyphs arranged over his effigy in the middle of the page encodes the omens for this particular set of Venus motions. One set, for example, reads:

I HE IS SEEN	2 IN THE EAST	7 WOE TO THE MOON	8 WOE TO MAN
3 GOD (SHOWN AT MIDDLE)	4 VENUS	9 THE DISEASE	10 OF THE SECOND MAIZE CROP
5 GOD (AT BOTTOM)	6 IS HIS VICTIM	11 WOE TO THE MAIZE GOD	12 WOE TO NIGHT

There is a ring of fatalism in these statements, a kind of resignation to the forces of nature that seems to pervade Maya thought. Basically these unfortunate tidings inform the user of the table that once Venus reappears in the east, the god of maize or night or certain diseases will be "his victim." In other words, bad crops, pestilence, or some other misfortune may transpire. Although the translations of all the omen-bearing hieroglyphs are not yet complete, it seems apparent that to the Maya intellect the ills cast down upon the world were aligned with sets of deities whose celestial behavior differed depending on the occasion. Knowing what these gods were up to was of paramount importance. Performing ritual sacrifices to them on the appropriate occasion was part of honoring a reciprocal contract. The gods help and provide for us and we pay them a debt—we offer them due compensation. Together we keep the universe in equilibrium.

The Venus omen text then moves ahead to the block of four Venus dates on the next full page, the second page of the table, continuing with the day name on line 10: *And then on 9 Ahau....* The eye moves horizontally to the conclusion of the full five-page run along line 10, encountering each set of five pictures as it goes. Next, the priest's eyes would pass back to the first page for the next run across line 11, then lines 12 and 13. Approaching the sixty-fifth 584-day cycle (5 pages × 13 lines × 584 days per page), the diviner would make an appropriate correction indicated on the user's page (not shown) that precedes the table. This would restore any temporal slack between the average Venus period of 584 days given in the table and the true Venus period as seen in the sky, which is a fraction of a day shorter than 584. Only then could the table's user recycle back to line I of the first page (which unfortunately is effaced by water marks in Figure 4.8) and thus resume another century's worth of Venus predicting—to an accuracy of one day in five centuries!

In the lower section of each page of the table, there are matching sets of "base dates" in the 365-day calendar that can be selected and meshed with those from the 260-day calendar above to give the complete calendar round date of a Venus appearance or disappearance. It is in the selection of these dates that we are offered a clue to an unanticipated rhythmic lunar beat that underlies the structure of the table. The three sets of base dates in the 365-day calendar are separated by time intervals that are divisible almost exactly by lunar eclipse cycles. This means that the Venus Table can actually be used to predict eclipses. In fact, as I have shown elsewhere, a lunar eclipse was visible on each lunar phase cycle immediately preceding each of the base dates in the table.[15] Evidently the cycle-minded Maya astronomers had discovered one of their gods' top secrets—how eclipses and Venus's appearances fit together in a harmonious, predictable way. Here was a pattern of celestial order unknown to their predecessors, even perhaps to their Old World counterparts. Our culture has tuned into a different Venusian rhythm. We insist upon seeing Venus marching in time to a solar beat, the way it appears in our current almanacs and planetary tables, which depict the rising and setting times for the planets according to the day and date of the year. Indeed, the solar year has been the principal baseline unit in our calendar since the Julian calendar reform during the time of the Roman empire (45 B.C.).

To understand Maya Venus astronomy, we need to inquire: What would a system of long-term Venus time reckoning look like if the moon, rather than the sun, were the principal indicator? If the lunar *month* rather than the solar *year* were the base interval, we can anticipate how a table of prediction of Venus events might appear. For example, a tabulation of Venus's motions might be expected to reveal multiples of the month of the lunar phases (29.5306 days) fitted together with the Venus synodic period (583.92 days). Remember the way our ancestors confronted the problem of the inexact fit between the number of phase cycles of the moon in a year of the seasons? Sometimes they added an extra month to the year. If the Maya culture that devised the Venus calendar also had struggled with the imperfect fit between lunar synodic month (29.5306 days) and seasonal year (365.2422 days), then we might further anticipate that a favored interval consisting of a whole number exactly divisible by all three key periods would appear in the Venus calendar.

The number 2,920 is an excellent candidate, because it *is* a whole measure of all three cycles—8 years of the sun, 99 months of the moon, and 5 cycles of the planet Venus. Recall that it also equals the time it takes Venus to trace out the five unique shapes of its track in the twilight sky during a sequence of morning or evening star periods. (See Figure 2.14.) In simple terms, five Venus rounds is about two days short of eight years and about four days shy of a whole number (99) of full moons.[16] Incidentally, although we cannot be sure whether they integrated Venus into it, the Greeks, who were also consummate cycle builders, knew this 2,920-day period very well; they called it the *Octaeteris* (connoting eightfold or octal). It was one of the earliest long cycles (about fifth century B.C.) used by the Athenians to fit moon phases, by which they reckoned days and months, into seasons of the solar year. Greek religion insisted upon carrying out the rites to the gods exactly on schedule, lest the member of the pantheon being addressed be displeased.

If the Maya Venus calendar uses the five visible Venus periods, a 2,920-day interval, as part of a lunar rather than a solar time base, might we also expect that Maya astronomers would have established the four components of each 584-day Venus cycle in time units based on the moon? Recall that the actual morning and evening star intervals that make up the Venus cycle pictured in Figure 4.8 average out to 263 days apiece, and that these were sandwiched in between unequal periods of absence, 8 and 50 days long on the average. If a lunar yardstick were incorporated in the calendar, then we might anticipate some sort of statement suggesting that each Venus appearance period is about nine months long, for 9 lunar synodic months fall just 2 days short of 263 days. But how would the 50-day period, which is not close to any recognizable lunar or solar interval, or the unwieldy 8-day period be expressed? And what about all the other periodicities, which we might *not* tend to associate with the moon or Venus in our way of thinking, that *could* appear in such a calendar? Our game of temporal hide-and-seek in unde-coded ancient documents could quickly degenerate into a random search.

Still, I suspect, cycle-minded lunar calendar builders would have been on the lookout for rhythms tucked away in the natural environment that could, with some ingenuity, be manipulated to fit together, the way we periodically pad our seasonal years via intercalation into sequences of 365, 365, 365, 366, to fit nature's seasons over the long run. Like the segments of the Venus cycle, no one of these intervals, taken by itself, is a precise approximation to the true length of the year, yet averaged together in long temporal chains, they grow into exceedingly accurate indicators of the sun's real whereabouts.

Looking at the bottom line of each of the five pages of the Venus Table, we begin to appreciate even further the lengths to which Maya astronomers were willing to go to distort short-term Venus time from reality so that, in the long run, its cycles would fit perfectly with the lunar measure they believed lay at the heart of their timekeeping system. Once again, "moon numbers" lurk in the four Venus intervals that make up each 584-day cycle. Remember that three of the four subintervals are only very rough approximations of the actual durations of appearance and disappearance of the planet. The real intervals must have been altered deliberately to correspond to lunar and possibly other ritual dictates not yet known to us. Also, recall that these intervals are approximately equivalent to multiples and half-multiples of the lunar synodic month. Thus, the morning star interval of 236 days is just a quarter-day short of eight months, and the 90-day disappearance period lies close to three months, while the 250 days the astronomers officially assigned to the evening star period is about a day and a quarter less than eight and a half lunar months. (The sole exception is the eight-day mean interval of disappearance before morning heliacal rise.)

What can these aberrant intervals mean in practice and what might the astronomers have had in mind when they used them? Suppose the moon were at a particular phase when Venus made its last appearance as morning star; then, after the first interval given in the table (236 days), the astronomer would know that the moon would necessarily be in that same phase and Venus would be about to make a first morning appearance. After the next interval (90 days) the moon would still be in the same phase and Venus would be scheduled to make its first evening star appearance. Following the third interval (250 days), the astronomer would know that just as Venus was about to make its last evening disappearance, the moon would be in the opposite phase; in other words, if it had been first quarter it would now have changed to last quarter. Finally, irrespective of the lunar phase, the last interval would count the eight days on the average until Venus would reappear in the morning sky. Thus the Venus moon reckoning system is a kind of astronomer's shorthand, a mnemonic for tracking the planet that probably derived from an earlier calendar based on the moon's phases rather than on the seasons and the solar year.

If the Maya Venus moon calendar seems a bit strange, think for a moment about the constraints in our own timekeeping system: We celebrate many holidays on a Monday in order to lengthen our weekend. Thanksgiving is always the fourth Thursday of November, yet the Fourth of July remains the fourth. Like the architects of our civic calendar, Maya timekeepers had limitations set upon them by the oft-conflicting demands of religion along with the desire to follow nature's metronomic beat. Thus, the 260-day cycle was concerned basically with ritual and divinatory schedules, while the 584-day Venus period was a cycle that consisted of timing a set of phenomena that occurred in the sky environment. We can understand the plight of the ancient Maya chronologist if we imagine that instead of scheduling a fireworks display on July 4, we were required to time the anniversary of America's independence with a natural celestial spectacle, such as a meteor shower. Clearly we would need to set up new timekeeping rules and modify old symbols—perhaps counting, according to some fixed prescription, forward to the nearest expected or backward to the most recently observed meteor shower.

Yale University linguist Floyd Lounsbury has given the best evidence for the actual date of the installation of the Dresden Venus counting scheme in real time: He argues that it was November 20, 934 A.D. or 10.5.6.4.0 days in the Long Count measured from the putative Maya creation. Thus, in vigesimal time count:

$$
\begin{aligned}
10 \times 144{,}000 &= 1{,}440{,}000 \\
5 \times 7{,}200 &= 36{,}000 \\
6 \times 360 &= 2{,}160 \\
4 \times 20 &= 80 \\
0 \times 1 &= 0 \\
\hline
\text{Total} &\quad 1{,}478{,}240 \text{ days,}
\end{aligned}
$$

or 4,047.45 years since creation. The Long Count appears very frequently on carved monuments that delineate real and mythical histories of the ruling dynasties of Maya cities, the idea being to link royalty to deity. It occurs less frequently in the codices, but when it is found, a Long Count date can offer a link between tabulated and real time.

Lounsbury chose the date 10.5.6.4.0 because it turns out to fall not only close to a first heliacal rise of the planet but also on the day name *Ahau*, in the 260-day cycle, which is traditionally associated with Venus as a lucky day. His discovery gives us a unique opportunity to test the accuracy of the Venus Table by comparing its predictions with modern back-calculations of actual Venus appearances in the sky. Lounsbury believes that by the tenth century, astronomers had already observed Venus's wanderings closely for hundreds of cycles. They must have become aware of the visible shortfall between the real Venus period of 583.92 days and the tabular version of 584 days they had been using. As it lost 0.08 day against every tabulated Venus period, the "great star" would have appeared about a day earlier every twelve cycles (about 20 years), or 5 days over the 104-year (Great Cycle) length of the table. As the user's page (page 24, not shown) prescribes, the Maya timekeepers remedied the discrepancy either by knocking off 4 days or by assessing a double correction of 8 days every time they reached the end of one or two Great Cycles in the table, respectively. The choice of multiples of four days had the added advantage of preserving the correct lucky name day for all future starting events in the table. (To see just how accurately the table follows Venus see the exercise in Appendix A.)

Because the Dresden document is an updated copy that incorporates data from about two centuries later than the 10.5.6.4.0 date, we can imagine the problem Maya skywatchers had been confronting for years as they attempted to revise earlier versions of the Venus calendar. Above all, they needed a reliable methodology for anticipating the all-important reappearance of Venus in the eastern predawn sky after its brief absence from view. Only then could the omens Kukulcan brings with him make any sense. The dilemma they faced is not so different from the one confronting Western timekeepers from the Roman empire to the Renaissance that ultimately led to the Julian and Gregorian calendar reforms (as a result of which we acquired our intercalation scheme of adding an extra day to certain years of our calendar). In both the Maya and Western calendar revisions, a day shift (in our case restoring the date of the spring equinox to the appropriate calendar date) was accompanied by the design of a set of rules to reduce the drift between real and canonized time. The only difference is that Julius Caesar and Pope Gregory were addressing the conflict between solar and seasonal time (for Gregory, the motivation was to determine when to celebrate Easter Sunday without having it fall on the Hebrew Passover), whereas the Maya cast of celestial characters—Venus, the sun, and the moon—played on a different stage.

There are other astronomical references in the Dresden Codex. A Mars Table that tracks the retrograde motion of that planet immediately precedes the Venus Table, and several of Dresden's later pages are given over to a seasonal table that incorporates eclipses and planetary conjunctions. A segment of the Paris Codex follows the passage of the moon among the thirteen constellations of the Maya zodiac. For a long time most of the rest of the codical pages were thought to be given over to endlessly repetitive 260-day cycles. However, as decipherment proceeds, scholars begin to realize that almanacs in the codices comprise ritual, astronomical, and calendrical cycles. They emerge as adjustable calendars that highlight events in real time, just as each edition of a modern farmers' almanac needs to be updated annually.

Of the more than 150 almanacs in the four Maya books, the Venus Table in the Dresden Codex fascinates us with its precision. It emerges as a splendid remnant of ancient Maya mathematical and scientific prowess, a hallmark descended from a golden age of Maya intellectual achievement. Like Stonehenge, the Venus pages entrap us, for when we probe their remains in depth, the Maya seem a bit closer to us. We begin to feel that they might have shared modern science's precise, quantitative way of comprehending the natural world. Unlike megalithic astronomy, the Maya written record gives us so much more to look for when we examine other kinds of evidence. Yet we must be prepared to believe that, even with the codices, so much of ancient Maya astronomy is still unknown.

How many unanticipated numbers like those wildly distorted Venus appearance intervals still await us? And finally, when we address the issue of how these people used their astronomical knowledge (the subject of the next section of this chapter) by pausing to read the omens attached to the numbers in the table, then we truly become sobered by the significant differences between our astronomy and theirs.

MAYA VENUS STAR WARS

The Aztecs were a people with a mission—they needed to keep the universe going. Believing themselves to be allied with the sun god, they waged a continuous battle against the forces of darkness, seeking to provide him with the precious liquid derived from the bodies of sacrificial victims that would propel him on his way. To avert cosmic disaster, the Aztecs waged constant warfare against the communities surrounding their capital city of Tenochtitlan. There they attained their supply of human hearts to fuel their light-bearing deity. It all goes back to the creation of the world by the gods of Teotihuacan who threw themselves into the cosmic fire to beckon the sun to rise, and to the man-god Quetzalcoatl. He was the one who fashioned the first humans from the ground-up precious bones of those who had lived in previous creations, cementing them together with blood shed from his member.

> And there [at Teotihuacan] all the people raised pyramids for the sun and for the moon; then they made many small pyramids where offerings were made. And when the rulers died they buried them there. Then they built a pyramid over them. The pyramids now stand like small mountains, though made by hand.... All were worshipped as gods when they died; some became the sun, some the moon....[17]

Thus did the chosen people, the Aztecs of Mexico, call themselves *macehualtin*, "those who are deserved by penitence"—or "humanity." Words and images of warfare alongside objects and events in the sky—this is the nature of Aztec militaristic cosmology. It may seem incongruent to us, but its roots can be traced in Mesoamerica all the way back to Teotihuacan, and in the Maya world it reached one of its high points.

I have focused so intently on the Maya worship of the planet Venus for two reasons: first, to give an idea of the depth of Maya astronomical calculation and prediction, and second, to offer clues for finding representations of the planet in the unwritten record. Given the importance the Maya accorded it, we ought not be surprised to discover Venus imagery turning up in sculpture, statuary, and mural paintings all over the Maya area and beyond. In these contexts we discover that one of the primary directives of Venus watching concerned the conduct of war. An eye-and-ray motif—consisting of a lidded eye in the shape of a circle framed by rays—has been identified in central Mexican codices and wall paintings as a celestial object, sometimes Venus.[18] This is because it often appears as an adornment of Kukulcan, the morning star god of the Dresden Venus Table, called Tlahuizcalpantecuhtli in central Mexico. We glimpse the same Venus star sign on the Platform of Venus at the ruins of Chichén Itzá, where it ties together a year bundle consisting of five strands of reeds and eight dots, the familiar commensuration of five Venus appearances or sky paths and eight solar years (Figure 4.10a).

The connection between Venus imagery and war was forged in Floyd Lounsbury's study of Long Count dates associated with a battle scene and accompanying associated events portrayed in a mural painting on the walls of a temple at the ruins of Bonampak (see Figure 4.10b).[19] That discovery was a major turning point in our understanding of the purposes of Maya astronomy.

While filming local native customs for the United Fruit Company in 1946, explorer Gales Healy, guided by a Lacandon Maya Indian, became the first outsider to enter the three-room temple at Bonampak, which is located in the rain forest of Chiapas, more than one hundred miles up the Usumacinta River from the ruins of Palenque. Despite the slime and calcareous encrustment that hid much of their detail from view, Healy recognized the paintings that covered the inner walls to be the finest Maya artwork the world had ever seen. But he was even more awestruck by what the pictures portrayed, for the delicately rendered composition depicted the pageantry attending the coronation of a ruler alongside scenes of warfare and brutality attending a great battle. Decapitated warriors lay sprawled next to the penitential vanquished, who lay slumped over and shed drops of blood from their limbs. Bonampak suddenly made the Maya seem more human. The paintings extinguished any prevailing thoughts about the Maya being a purely philosophical, placid people in search of eternal truth, with little interest in self-glorification or imperial conquest.

How did Bonampak's temple change our view of Maya astronomy? In a sky panel above the battle scene, Venus hieroglyphs ride on the backs of

(a)

Figure 4.10 Venus symbolism: (a) Venus symbol (left) with year sign
(top) and bundle of years, with 8 dots (right), from the platform of
Venus at Chichén Itzá.

weird animal and human creatures that represent signs of the constellations
of a Maya zodiac used to chart the course of that planet along the stellar
roadway.[20] Moreover, Venus events have been identified with dates given in
the associated inscriptions on carved monuments at Bonampak. Lounsbury
was able to show that these dates match actual morning heliacal risings as
seen from that place in the eighth century A.D., about the time the paintings
were made. That Maya rulers often are depicted wearing the Venus sign as
part of their military costume lends support to the interpretation that this
planet has something to do with warfare. But what does Venus the dart-
thrower bring in his wake? Like the omens written between the pictures of
the Dresden Codex, the answer lies in the subject matter of the murals. We
still do not know enough about the written and pictorial record from ancient

(b)

(c)

Figure 4.10 (*Continued*). (*b*) War scene from mural painting from Bonampak. Note the zodiacal band of four constellations at the top, which includes the peccaries and the tortoise (compare Figure 2.15b). (*c*) Venus's glyphs under eyes of a rain god mask at Chichén Itzá. Compare these with the Venus hieroglyphs in Figure 4.9.

Mexico to understand why, of all planets, Venus was chosen to be the one associated with the conduct of war ritual and the initiation of battles.

As I indicated earlier, Maya Venus war symbolism is not restricted to Yucatán. At the ruins of Cacaxtla, atop a huge fortress in highland Mexico far from Yucatán, archaeologists working in the mid-1980s unearthed a painting of a Venus-skirted warrior and his female companion. The man holds a Venus symbol and he is adorned with a scorpion's tail. Another testimony to the widespread connection between the motions of Venus and the conduct of war is found in an adjacent room excavated in 1975. It is another battle scene like the one depicted in the Bonampak ruins. The winners, who stand gloating over the nearly naked bloodied warriors lying crumpled and vanquished in their wake, wear Venus star-studded robes. These and other accoutrements have been likened to similar garments worn at Teotihuacan. The "star warriors" stand against a border made up of half five-pointed stars. As intimidating and as radiant as Kukulcan of the Venus Table, these soldiers may be thought to be as dangerous and warlike as their patron god, ready to bring down their wrath upon the backs of the enemy. In still another room, the lords of Cacaxtla appear to be executing the surviving captives as an offering to the gods of fertility, perhaps in order to ensure that water and new maize will not be in scant supply in the coming season.

The cult of Venus regulated warfare and sacrifice and its connection to fertility spread across Mesoamerica. Mayanist John Carlson has traced this same scene back to similar "Venus enclosures" at Teotihuacan. In one of them a green storm god holds a staff of lightning in one hand and an offering of blood droplets in the other. In another, goggle-eyed rain/fertility gods display dripping hearts on their hooked knives. Carlson believes that these scenes record actual heart sacrifices by priests impersonating the sky gods as part of a choreographed procession during which blood was offered directly to the appearing or disappearing Venus.[21]

The record written in the codices and the icons in mural painting seem to tell related stories. The Dresden Codex deals principally with *perception*, how precisely the astronomers tracked the planet and the Venus phenomena with which these people were concerned. The plastic arts deal more with *conception*, or where Venus fit into the Maya worldview. At least for a time he was the patron deity of warfare. In the architectural remains, we again find astronomy and ritual intertwined; their temple stairways to the stars were built for the purpose of religious worship. Here is a theme that resonates with our interpretation of Stonehenge as an astronomical observatory as well as a place of assembly.

ASTRONOMY
AND MAYA ARCHITECTURE

The accomplishments of the Maya astronomers revealed in their written legacy appear even more dazzling to our modern mentality when we realize that they discovered all their intricate celestial cycles without the aid of the instruments of precision that so greatly assisted the Old World astronomers who developed our own modern calendar. New World astronomers were not totally devoid of technology. We know they employed narrow tubes and windows, which they incorporated into the architectural plans of many of their buildings expressly for the purpose of making celestial observations. As we have seen, notched sticks also may have been set up in the doorways of some of the buildings to register horizon events, but these tools differ markedly from the graduated angle-measuring devices of the European astronomers, such as the sextant or telescope mounting.

As we approach Maya architecture from an astronomical perspective, we again need to remind ourselves that ancient science was closely allied with religion. It was not the socially detached enterprise many of us make it out to be. The Maya religious domain consisted of temple pyramids and ceremonial centers. These places were the center of their material existence and dominated their lives. Heaven was the realm of gods and ancestors whose behavior was epitomized by the natural order of the cosmos. But heaven and earth are necessarily connected because humans beings, cast in the image of their gods, are guided by heavenly forces. Historian of religion Paul Wheatley has argued that those religions that specifically associate the creation of the universe with the origin of humankind usually dramatize the cosmogony by attempting to reproduce on earth a miniaturized version of the cosmos. Maya centers of worship became replicas of the ruler's concept of the dwellings of deities and dead ancestors. For example, the penetration of the forecourt through the sanctuary to the inner shrine on a two-dimensional surface might be equivalent to the ascent of the ruler to progressively higher levels of heaven in the third, or vertical, dimension. Not only the temples but also the rituals dramatized within them connect the imperfect earthly realm with that of the divine.

We who worship inside churches, cathedrals, synagogues, or mosques associate religious activities with interior space. For the Maya, living in the humid tropics, the inside of an enclosure would be the last choice for a venue in which to celebrate communion with the divine. I think the evolution of Maya architecture gravitated toward minimalizing interior space in favor of large massive pyramids that surrounded huge plazas at least in part for this specific reason. It is likely that the elaborate stagings that attended sacrificial rituals were conducted in the open spaces that fronted Maya ceremonial buildings. Reviewing stands often project outwards from temple doorways, thus providing a logical place from which a royal bloodletting could be viewed by people positioned below.

Any building possessing a peculiar shape or orientation relative to other buildings at an archaeological site immediately arouses our astronomical suspicions. Could architects deliberately have shifted a wall or a doorway out of place in order to fix a sighting post to register a special celestial event? One of ancient America's most peculiar buildings can be found at the ruins of Chichén Itzá, Yucatán's major tourist attraction.

After the collapse of the Maya cities of the southern Yucatán peninsula about the tenth century A.D., Chichén Itzá (it means the mouth of the well of the Itzá, one of the last people to settle it before Spanish contact), along with Uxmal and later Mayapan, grew into one of the post-Classic northern centers of power. Chichén then housed a hybrid culture that synthesized the styles and ideologies of old Maya ways together with an infusion of the Nahua-speaking, highly militaristic Toltec culture, which had expanded out of the highlands of central Mexico. Iconography there is rife with jaguars and eagles devouring human hearts along with the same feathered-serpent symbolism traceable all the way back to Teotihuacan. This was the culture that built the temple of Kukulcan with its four grand stairways capped by balustrades, down which the equinox sky serpent still descends (see box on page 145); the Temple of the Warriors surrounded by colonnaded hallways; and Mesoamerica's largest ballcourt, nearly 100 yards long, with its two stone rings positioned 27 feet high on surrounding vertical walls; and, of course, the circular 190-foot-wide, 130-foot-deep half-water–filled sinkhole (the "Well of Sacrifice") that gave the site the first half of its name.

The Caracol tower of Chichén Itzá (Figure 4.11a) is surely the most famous example of a cockeyed building. So beguiling are the Caracol's asymmetries that archaeologist Sir Eric Thompson once remarked that "something must have been wrong with the architectural taste of the people who built it!"[22] The Caracol's lack of aesthetic appeal (at least to some) is largely a result of its odd shape as well as its skew. This has led some investigators to suggest a functional motivation for this aspect of its design. It has been called the gnomon (a vertical shadow-casting device) of a huge sundial as well as a military watchtower, but of all the uses proposed, astronomical observations seem most successful in accounting for the peculiarities of its situation and orientation.

When I first visited the Caracol in 1970, the narrow shaftlike windows immediately attracted my attention. Archaeologists in the 1930s who undertook the first thorough excavations there had measured several alignments taken along the jambs. Being unacquainted with tropical skies, they sent their data for analysis to the U.S. government's Department of Terrestrial Magnetism, which shot back the opinion that some of the orientations may have been intended to mark lunar standstills. This made little sense, for there had been no record of any alignments on other buildings in Mesoamerican building alignments pointing in these directions. Moreover, the measurements we made in 1973–75 revealed angles that failed to match the lunistices. Despite its circular form, the Caracol clearly was not built with a Stonehenge-type astronomy in mind.

The lower platform (Figure 4.11b), the first building unit of the Caracol, was constructed by the Maya about A.D. 800. It is a rectangular area about 3,500 yards square. The large front stairway faces 27½° north of west, which puts it conspicuously out of line with the other buildings at the site. The sunset position at summer solstice lies within 2° of this direction, but an even closer match is provided by the northern standstill of Venus. Recall from our extended discussion of the motion of Venus in Chapter 2 that the planet will arrive at a particular horizon extreme every eight years

(a)

Figure 4.11 The Caracol of Chichén Itzá. (a) Asymmetries seem to indicate that the purpose of the building was to record significant astronomical events.

and that its length of disappearance, which averages eight days, also varies with the seasons. Marking the Venus standstill could have proven useful to astronomers who wanted to predict when the morning star would arrive, and this information, as the Dresden Venus Table reveals, was their prime target.

Above the lower platform, embedded in the stairway of the upper platform, lies a niche containing an altar mounted by a pair of columns. The altar is aligned asymmetrically relative to the upper platform and it, too, points to the northern Venus extreme. The columns retain flecks of black and red paint. Red is the color associated with the place of appearance of Venus in the Dresden Table and black usually stands for west. Is it possible that the painted columns served as a monument to Venus in the east as morning and in the west as evening star? (Recall that *chak ek*, the common Maya name for morning star, means red star or great star.)

To ascend to the top of the Caracol tower you need to crawl through a snaillike passageway that gives the Caracol its modern name (*caracol* is the Spanish word for snail). Up in the turret astronomers would have been able to make two more Venus observations (Figure 4.11c). Three horizontal shafts emanate from a small rectangular chamber and look out onto the flat

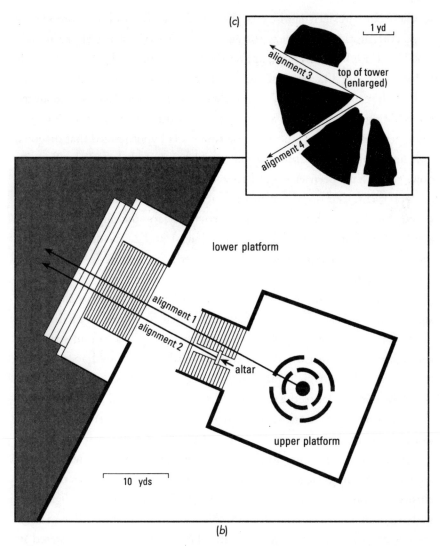

Figure 4.11 (*Continued*). (b) In the plan of the whole building, two of the alignments to the Venus northern standstill of about A.D. 1000 are shown; the three circles show the openings in the ground floor of the tower. In (c) what remains of the top story of the tower is shown. Alignment 3 also points to the Venus northern standstill; alignment 4 indicates the southern standstill.

southern and western landscapes. While the largest will accommodate a person attempting to squeeze through, the other two shafts are so narrow that they leave little doubt regarding their function. Surely they were made to look through. The window jambs frame short segments of the southern and southwestern horizon. When archaeologist Oliver Ricketson analyzed the building for astronomical orientations in the 1920s, he was drawn to these peculiar windows built into the tower. He hypothesized that diagonal sightlines, for example from the inside right to outside left jamb of a window, could have been employed to accurately pinpoint the position of a horizon event. Our own later measurements (made in the mid-1970s) lend strong support to this hypothesis, particularly for the alignments of the inside left to outside right jambs of windows 1 and 2. We determined that these directions precisely marked the northerly and southerly standstills of Venus along the horizon.[23]

Venus may have more to do with Caracol architecture than alignments alone would indicate. Early historians tell us that Kukulcan, when disguised in the form of the wind god, Ehecatl, was worshipped at round structures throughout Mesoamerica. Moreover, the Dresden Codex with its accompanying Venus Table, which traces the celestial course of Kukulcan, probably was drawn up not far to the east of Chichén Itzá. It is even possible that the astronomical observations delineated in this document were collected by astronomers in this very tower. In view of all the evidence, it seems surprising that an intimate connection between Venus and the Caracol was not realized until relatively recently—within the past two decades. Perhaps the elusive motion of the planet combined with a failure to connect alignments with written calendars contributed somewhat to its neglect by modern investigators. Nonetheless, the ancient Maya seemed to have mastered its course quite well.

Another disoriented building in ancient Mayaland also may have been a Venus observatory. Uxmal, one of the largest sites in northern Yucatán (and famous for the elegant simplicity and precision of its architecture), shared the political platform with Chichén Itzá in the late ninth and early tenth centuries A.D. Practically all of its buildings line up along a 9° to 10° east of north axis, but the House of the Governor, an acknowledged masterpiece of Maya architecture, is a notable deviant from the plan, which is pictured in Figure 4.12. Art historian Jeff Karl Kowalski, who has completed a detailed analysis of Uxmal's inscriptions and iconography, believes the building was the residence of a divinely inspired ninth-century ruler. Kowalski's idea does not preclude the possibility that the House of the Governor also was a place where commoners came to worship their king and perhaps to witness him performing his penitential sacrifice. The sculpted frieze above the doorway consists of a two-headed serpent bar, his instrument of office, upon which signs of the Maya zodiac are inscribed. An effigy carved on a nearby stela depicts a ruler standing on a jaguar throne that looks just like the one fronting the temple (in Figure 4.12b). That the ruler is "exalted, seated in majesty within a cosmic frame and associated with dozens of rain god masks [indicates that] he governs with supernatural sanction," writes Kowalski.[24] The king's effigy looks out eastward from its perch exuding authority over all the surrounding cities scattered across the flat landscape.

Constructed on a large artificial platform about 220 yards on a side, the Palace of the Governor deviates in a clockwise direction by 15° from the main axis of the site. The building faces 28° south of east. Moreover, unlike the other buildings, its central doorway faces outward toward the horizon rather than inward toward the main axis of the site, as is customary in the architecture of this region. In 1973 my colleague, the German-Mexican architect Horst Hartung, and I were standing on the porch fronting the House of the Governor when he noticed that the view from the middle of its thirteen doorways seemed perfectly aligned with a rather prominent bump on an otherwise featureless horizon. The view through a telephoto lens (Figure 4.12b), and later the telescope of our surveyor's transit, revealed a pyramidal shape very heavily overgrown with trees shimmering in the midday haze; we estimated its distance at 2 or 3 miles. Our first attempts to get to the promontory were thwarted, for no road and scarcely a pathway existed. On our third foray we obtained a couple of four-wheel-drive vehicles and elicited the help of a local guide equipped with machete and rifle, and we hacked our way from the Pan American Highway through the scrub vegetation to a rather impressive 50-foot-high building surrounded by several lesser structures flanking a huge open plaza.

Our measurements revealed that the central doorway of the Governor's House lines up precisely with the largest structure at the neighboring site, now known to be Cehtzuc. To our surprise, this direction also coincided with the southernmost standstill of Venus rising.[25] The Venus connection with the

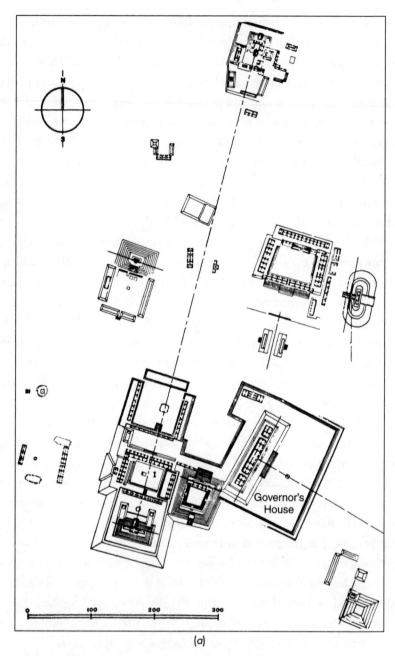

(a)

Figure 4.12 (a) House of the Governor, Uxmal, deviates from the average alignment of all the other structures at Uxmal (long axis). It points outward to Venus's southern standstill over a pyramid in a nearby city.

(b)

Figure 4.12 (*Continued*). (b) Telephoto view of the Venus alignment from central doorway.

orientation of the Palace of the Governor is supported by information derived from the iconography that we find on the building (Figure 4.10c). In addition to the zodiac mentioned above, more than 350 Venus hieroglyphic symbols are positioned under the eyes of an array of rain god masks that adorn the upper facade of the building, the very same symbols that repeatedly appear on pages of the Venus Table in the Dresden Codex. Also, a possible representation in Maya dot-bar notation of the mean disappearance interval of Venus before morning heliacal rise (the crucial 8-day entry in the Venus Table) appears sculptured over the eyes of Chac (rain god) masks on the northeastern and northwestern corners of the uppermost platform of the building (Figure 4.10c). Recall that according to the Venus Table it is precisely after the disappearance interval of 8 days that one would expect to see Venus rising in the east.[26] These observations may be of special significance because the Chac masks on the facade of the Palace of the Governor are the only ones at Uxmal that bear the Venus hieroglyph as a decorative element.

To judge from the vast quantity of imagery showing rulers in exultation all over Yucatán, the classic Maya seem to have devoted a good deal of time and effort to ceremonial ritual. For example, on Stela B, the accession monument of a ruler of eighth-century Copan, King 18 Rabbit (Figure 4.13) wears the symbols of power: a belt of ancestors' heads, bloodletting lancets, sky birds, corn foliage, and an image on his loincloth of the sacred tree that joins sky and earth. At the top of the stela a scene depicts him communing with his sky ancestors positioned atop clouds that surround a sacred mountain. We can imagine this king in real life surrounded by his deputies, who are weighted down by their heavy regalia as they impersonate the deities. Were they celebrating a victory in battle or his royal betrothal, or were they overseeing the ceremonial ballgame by attending the completion of one of the paramount time cycles that ensured the continuity of his dynasty? Perhaps they were commemorating his accession. The death of a ruler and the succession to office of his replacement—such an abrupt discontinuity, so unsettling, for the Maya would be like a kink in the circle of time. How would the people know that once their leader was gone, his son would possess the same powers? What guaranteed it? For Americans, continuity of rulership is promulgated by faith in the democratic process; for the Maya, it was by faith in the bloodline. Blood is the answer. Kinship and dynastic bloodlines are proclaimed the principle of immortality in so many Maya inscriptions.

Figure 4.13 King 18 Rabbit of Copan is depicted on the front of Stela B. He holds a sky serpent bar, an instrument of office, and has stingray bloodletters firmly in a pair of holsters held at the waist. Above his headdress he communes in a trancelike state with his ancestors perched atop Macawitz, the sacred mountain, amidst smoke scrolls.

Although our modern computers can regenerate with relative certainty the sky events that unfolded above the ancient Maya cities of Chichén Itzá and Uxmal, we can only speculate about what happened when Venus appeared over a special temple that was dedicated to a planetary god. Perhaps the ruler actually sat on his throne when his celestial ancestor made an appearance in the form of the resurrected Venus. Maybe he even spoke the appropriate words to influence the gods in their course of action as he performed a penitential bloodletting from his private parts. And the assembled throng looked on in awe.

Such elaborate theatrical stagings reveal Maya beliefs about the essence of heavenly power in a direct and forceful way. The goal of specialized Maya ceremonial architecture may have been to instill in the viewer-participant the same sort of passion that might well up in the breast of the medieval Christian pilgrim who saw the sun shine through the stained-glass windows of the cathedrals of Chartres, Reims, or Cologne for the first time, or today's patriot watching flags flying at a formal military occasion. For the ancient Maya, sky phenomena were laden with powerful messages.

While Venus orientations in Maya architecture make sense as resource material for what appears in the written form of the Maya calendar, we must be careful about overinterpreting ceremonial architecture. Some of today's popular literature offers numerous examples of Maya pyramids as direct pipelines that convey celestial messages to a throng in the midst of celebrating a ritual. "Staged" natural events, such as inaugurations, celebrations of victories in battle, great royal turning points, and so on, were very likely commemorated publicly in the open spaces fronting the pyramids on days when some important, directly observable celestial event occurred. Many contemporary scholars, perhaps mistakenly, tend to view the ancient Maya ruler as a super-shaman engaged in reading the future in his stars. Maybe this is because today so much Maya religion is conveyed to us through shamanistic rituals and practices. Others seriously question the assumption that there is a single worldview that has permeated Maya culture continuously from the present all the way back to its classical heyday in the eighth century. Are we getting the Maya astronomy we desire and deserve?

Given its complexity and the hieroglyphic code in which it is stated, it is likely that only royal sky specialists would have had access to the knowledge gleaned from observed timings of Venus marked by a distant pyramid or viewed through a narrow window. Only their actions could enable celestial predictions and the attending celebration and reaffirmation of a celestially based myth to proceed. In this sense the Governor's Palace, like the Caracol tower and many other Maya buildings, can be regarded as an observatory, a device for penetrating the minds of the gods. Planetary guideposts revealed themselves to be the true source of power and authority, the guarantors of the legitimacy of change in a belief system that penetrated both the public and the private sectors of Maya society. By examining written documents along with the architectural alignments that back them up, we learn that Maya astronomy was not *just* astrology pure and simple. We discover that their sky observations were decidedly different from our own, not only in the way they gathered their data, but also in the uses to which they put them. As with Stonehenge, the Maya world offers us an astronomical science inextricably linked to religious pursuits, a combination we will see repeated once again, with a heavy nod to statecrafting, when we look south to the land of the Inca.

THE SERPENT DESCENDS

The subtle influence of light and shadow in Mesoamerican architecture may be another way Maya people expressed their knowledge of astronomy. A *hierophany* is a showing of something sacred. It usually consists of a phenomenon in the land- and skyscape displayed via the deliberate arrangement of architecture. The idea is to provide a powerful religious experience that people who assembled in the ceremonial center could witness. One hierophany that has gained considerable notoriety in the Maya world is the equinox phenomenon that occurs at the Temple of Kukulcan (the Castillo) at Chichén Itzá (below). Today the event attracts tens of thousands of celebrants and other visitors from all over the world, thanks to the easy access of Chichén to tourists. First recognized by photographer Laura Gilpin in the 1940s, the phenomenon was described in detail by a Frenchman, Jean Jacques Rivard, a generation later in an obscure note.[27] The mid-1970s saw the event popularized, and by the 1980s it had attracted a worldwide following from Maya aficionados to New Agers. Here is what happens: Late on the afternoon of either equinox (and for a period of up to a week before and after), the nine tiers of Chichén's largest pyramid cast shadows on the western balustrade of the north stairway. The shadows form a wavy line that seems to attach to the stone serpent head situated at the base. Interestingly, the north stairway is the only one with serpent heads extant, and it

(continued)

(*Continued*)

also leads to the main doorway of the temple at the top. As the sun descends, the undulating shadow rim touches the top of the balustrade forming seven triangles of light, not unlike the triangular pattern on the back of a rattlesnake. Rivard went on to speculate that after the serpent descends the stairway he moves in the direction of the sacred well, the home of the rain god, to the north. He also indirectly ties other buildings representing Venus and the underworld into the hierophany.

CHAPTER FIVE

CITY AND COSMOS: ASTRONOMY AND THE INCA EMPIRE

Holy beings, who among you is not saying, "let it not rain, let it not freeze, let it hail"? Speak immediately.

—Ruler Tupac Inca speaking to his *huacas*

THE CORICANCHA AND ITS ASTRONOMICAL SYMBOLISM

"The Incas had little knowledge of astronomy and natural philosophy since they had no letters." So wrote sixteenth-century chronicler Garcilaso de la Vega, part Inca himself, in his *Royal Commentaries of the Inca*.[1] Yet the Inca were master architects who, by the eve of the Spanish conquest, had developed an empire stretching over all of western South America. And they achieved it all in less than a century! Actually, a rich tradition dating back to the cultures of Huari, Chimu, and Nasca served as the base for the development of Incaic ideas about cosmology. It may be premature to dismiss ancient Andean astronomy as unimportant just because that civilization exhibited no elaborate system of writing as we know it to preserve their calendar.

Who were the Inca? Just as the Maya had the Zapotecs and the Olmecs to build upon, so too the Inca culture rested upon the foundation of Huari and Tiahuanaco (Tiwanaku), the empires that immediately preceded them. By the twelfth century these once mighty states to the south were reduced to less tightly controlled competing communities by drought and ecological mismanagement. Out of their shambles, barely a century before Spanish contact, emerged a skilled organizer and military leader named Pachacuti Inca Yupanqui (Pachacuti means "earthquake"). A capable ruler, by the mid-fifteenth century Pachacuti managed to beat off all his competitors. Based in the valley of Cuzco, now established as the southern capital, he and his kin undertook a campaign that led, in just a few generations, to an empire that stretched from Quito (Ecuador), the northern capital, to Santiago (Chile)—a region 3,400 miles long and 385,000 miles square that encompassed twelve million people.

Just as Aztec legend claims descent from the remains of a once great culture, so too Inca legend tells of its people who were born of the sun god, Inti. He reigned at the city of Tiahuanaco, impressive for its stately stone sculpture and a famous semisubterranean temple, the capital and ceremonial center of an empire that once stretched around the southern reaches of Lake Titicaca, 300 miles southeast of Cuzco. The first Inca, four generations of brother and sister pairs, were said to have traveled underground, finally emerging at Pacaritambo, a mythical cave overlooking the place where the gods had instructed them to drive their staffs into fertile ground—the valley of Cuzco. But they could not claim the turf on which to extend the rule of their gods until they beat off the Chancas, invaders on the west. Thus it was that a ragtag band of settlers who had eked out a living on the slopes of the valley built legend into the history of one of the great empires of the New World.

Of all the reasons we can imagine for the success of the Inca empire, which lasted but a century before the Spaniards interrupted it, the one that must be taken most seriously is the strict order and the high degree of organization that was built into every component of it. An early Spanish visitor to the highland Inca capital remarked:

> Nowhere in the kingdom of Peru was there a city with the air of nobility that Cuzco possessed . . . compared with it, the other provinces of the Indies are mere settlements. Such towns lack design, order, or polity to command them, but *Cuzco* has distinction to a degree that those who founded it must have been people of great wealth.[2]

Thanks to the writings of Cieza de Leon and other Spanish chroniclers, Cuzco is one of the few places in the Americas where, even though only a small portion of its once stereotypical walls of exquisitely carved polygonal andesite blocks remain (see Figure 5.1), we nonetheless have some reliable data on how pre-Columbian cities were planned and organized. Ethnohistoric, ethnographic, and archaeological inquiries have demonstrated that Andean concepts of both time and space are inextricably bound to religious, social, and political organizing principles, all of these being embed-

Figure 5.1 Handworked stone blocks that make up the wall of the Coricancha of Cuzco still lie in place despite repeated destruction by earthquakes of the Church of Santo Domingo that the Spaniards erected on top of it. This temple, once decorated in gold, was the center of the *ceque* system.

ded in the structural plan of the city. The system that binds these elements together is one of the most unusual in the history of city planning, for Cuzco's basic layout is *radial,* an expressive form that seems very special and important in Andean society—a plan that incorporates Inca knowledge of the sky.

There is little doubt that the Inca watched the heavens closely and that they developed a precise calendar based on what they saw. Much of their sky watching likely took place in the Coricancha (literally "golden enclosure"), the Temple of the Ancestors, also called by the Spanish the Temple of the Sun. From the Spaniards' descriptions of it when they arrived in 1533 it must have been a marvelous structure.[3] It was covered all in gold with a huge golden sun disk facing the great luminary as it rose (thus its name). Historian John Hemming recounts one chronicler's description of the place and how the Spaniards looted it:

> "These buildings were sheathed with gold, in large plates, on the side where the sun rises, but on the side that was more shaded from the sun the gold in them was more debased. The Christians went to the buildings and with no aid from the Indians—who refused to help, saying that it was a building of the sun and they would die—the Christians decided to remove the ornament . . . with some copper crowbars. And so they did, as they themselves related." The Spaniards pried off seven hundred plates, which Xerez [the chronicler] reported as averaging some 4½ pounds of gold each when melted down. "The greater part of this consisted of plates like the boards of a chest, three or four palmos (2–2½ feet) in length. They had removed these from the walls of the buildings, and they had holes in them as if they had been nailed."[4]

The king was said to have sat upon a throne built into the east-facing wall, which was pierced with many precious stones and emeralds set in the holes. Other descriptions speak of tubs of gold, silver, and emeralds. Today, like so many important pre-Columbian buildings, the Coricancha is covered over by a Spanish church, the Church of Santo Domingo. But enough of it has been excavated by archaeologists to reveal that it consisted of a rectangular enclosure (*cancha*) fronted by four buildings of cut stone with thatched roofs. The fact that gold (*qori*) adorned the building has given rise to its name, the "Golden Enclosure." It must have been a truly important edifice, on a par with the Church of the Holy Sepulchre in Jerusalem.

The Coricancha was also dedicated to the most important heavenly bodies worshipped by the Inca: the sun, the moon, Venus, and the Pleiades. Venus, attendant to the sun, was recognized as the same body whether it appeared ahead of or behind the sun. They observed its heliacal rising and setting. The planet was thought to have been ordered by the sun to go sometimes before him (as morning star) and sometimes behind (as evening star) but always to remain close by. According to Garcilaso:

> Another hall, next to that of the Moon, was dedicated to the planet Venus, the Seven Kids and all the other stars. The star Venus they called Chasca, meaning "having long curly hair." They honored it saying that it was the Sun's page, standing closest to him and sometimes preceding and sometimes following him. The Seven Kids they respected for their peculiar position and equality in size. They thought the stars were servants of the Moon and therefore gave them a hall next to that of their mistress, so that they would be on hand to serve her. They said that the stars accompanied the Moon in the sky, and not the Sun, because they are to be seen by night and not by day.[5]

A page from a manuscript (Figure 5.2), produced shortly after the conquest by the chronicler Joan Santa Cruz Pachacuti Yamqui, attests to the importance of Incaic constellations as well as to the central role of the Coricancha temple in astronomical endeavors. It also dramatizes two basic structural principles of Inca cosmology: dualism and vertical hierarchy, both of which persist today and figure prominently in Inca alignment schemes.

That all things should come in twos is not an idea unique to the Inca worldview. Day and night, winter and summer, male and female—these are a few of the entities we all experience that give rise to the idea that order in the world consists of paired opposites. This may be why cosmologies ranging from Maya to Mesopotamian attribute the origin of the world to dual

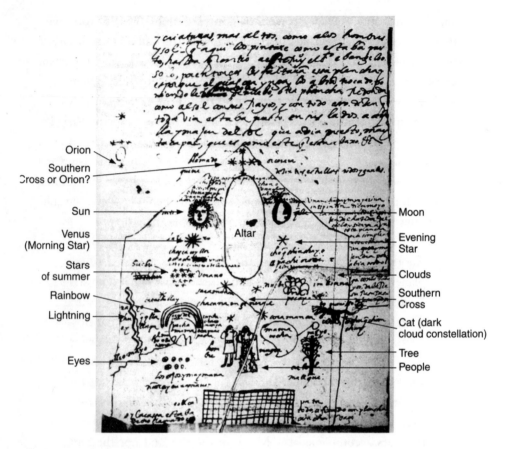

Figure 5.2 Diagram of the Inca Coricancha drawn by chronicler Joan Santa Cruz Pachacuti Yamqui, showing the celestial objects worshipped there.

creators like the Mesoamerican hero twins or the Chaldean personified deities who represent land and sky, fresh and salt water.[6] Even our modern model of the microcosm rests on the principles of positive and negative charge, and our "Big Bang" macrocosm began with the differentiation of the universe into matter and radiation. A second paragon of order more peculiar to those who live in a mountain environment consists of the "up and down" classification of things that make up the world. Anyone who takes a car trip from the coast to the Andean highlands can experience the extraordinary variation in Andean ecology in a single day. Within a few hours you rise from the foggy maritime environment of Lima, Peru's capital city, to frigid tundralike passes bordered by snow-topped mountains

18,000 feet high. Each climate zone along the ascent exhibits its own micro-climate and productivity: cotton fields, cactus, tree fruit, squash and fish-eries on the coast, to maize in the middle altitudes to ever hardier varieties of potato and related tubers as you rise to the elevated plateaus, or *puna*, where herders graze their llamas and alpacas. Anthropologist John Murra[7] has demonstrated that in Inca times, control of these vertically arranged economic zones was maintained by systematically relocating people and enacting strict laws dictating who should have access to farmland within each region. Even where they should live within that region was regulated by law—a clear indication of the extraordinary degree of organization that existed within the empire.

Pachacuti's diagram illustrates these two principles of organization—dual and vertical order—joined together. For example, the temple, located in the center of the Inca capital at the junction of two rivers, is shown in a lateral view, divided into solar (left) and lunar (right) houses. Venus as morning star lies below the sun. They called it *Chasca*, a word meaning "star" in the spoken Quechua language. The Pleiades and the Southern Cross, located opposite from one another in the sky, are among the con-stellations represented. The former is labeled the "stars of summer," prob-ably because in the latitude of Cuzco the Pleiades make their first appearance in the eastern sky opposite the setting sun shortly before the arrival of the December (summer) solstice in Cuzco. Thirteen Pleiades stars are shown, not an unrealistic number to be seen at this high altitude in the rarified Andean air. Below we see the Pleiades again, this time represented as the seven eyes of Viracocha, god of thunder and creation. The Southern Cross, called the "hearth" (with a fifth star added in the right place!), appears at the center of the diagram below the oval Coricancha altar. Another version of the Southern Cross may be represented by the group of stars near the rooftop of the Coricancha (it may also have been intended to represent a Christian cross). Its axis points directly to the apex of the roof of the building, which may be the south celestial pole. (Today mariners who sail the south seas still use the Southern Cross to find the south pole, as there is no bright star to mark the fixed pivot of sky motion in the South-ern Hemisphere.) In 1400 A.D. it pointed to within 5° of the pole. Orion is an alternative representation of the rooftop constellation, the three belt stars in the middle with bright Betelgeuse and Rigel forming the top and bottom extensions of the cross. Interestingly, an almost unmistakable Orion configuration is sketched in at the left margin of the page outside the Coricancha. The region of the Great Nebula in Orion is circled in precisely the correct place to support this identification (see Figure 2.15b for the way the Maya saw this constellation). A possible dark cloud constellation (*chuqui chinchay*—a puma) appears at the right of the diagram.

Pachacuti's picture captures the essence of the two foundation principles of duality and verticality in Andean cosmology. Cleaved vertically, the left side of the ledger seems to represent things masculine, while the right side pertains to phenomena of a feminine nature. For example, the cosmic cat symbolizes *pacha mama*, or "earth mother." Pachacuti's diagram also may be divided horizontally into three layers: the heavens (top), the real world (center), and the underworld (bottom). The lightning flash, or *rayo*, on the left is the beam of the sun. It descends from above at the time of the passage of the sun in the zenith. Indeed, planting is still geared to these events in the calendar.[8] This vertical dualism turns out to be the key to the doorway of cosmological expression in the architectural plan of ancient Cuzco. Thus, as in the case of the Maya and Stonehenge, our desire to understand ancient astronomy leads us directly into the heart of a people's civic and ceremonial space. As we shall see, astronomical knowledge is expressed via sight lines connecting the place of appearance of sunrise and sunset on important days in the calendar.

THE RAYLIKE ORGANIZATION OF CUZCO

Cuzco is situated in latitude 13⅓° south at the junction of two rivers in an 11,000-foot-high mountain valley in the central Andes. It is said to have been expressly planned and laid out by the Inca Pachacuti himself in the shape of a mighty puma. The city is protected on the north by the huge fortress of Sacsahuaman with its zigzag walls made of cubic-yard–sized, worked granite stone blocks; it represents the puma's head. The hindquarters and tail are styled by the canalized rivers, which are made to join in a *tinkuy*, the harmonious balancing and blending of opposites where things come together. The neighborhood of this place is still called Pumac Chu-pan, or the "puma's tail."

The Inca called their city *Tahuantinsuyu*, or Four-Quarters, because of its basic quadripartite plan. Though it boasted a population of more than 100,000 at its height and contained over 4,000 structures, records indicate that except on festive occasions, only the nobility and high-ranking visitors ever penetrated the city's center. Only they saw its exquisite masonry build-ings surmounting wide plazas, many of them displaying gold sheathing and housing carved stone effigies of the deities. Ringed by mighty snow-capped mountains, the view from the *ushnu*, or ceremonial platform, in the middle of the plaza must have made a lasting impression on dignitaries who came from afar, who were said to have entered this "navel of the world" laden with precious gifts for the emperor.

Cuzco was segmented spatially into halves called *Hanan* (upper, located in the northwest) Cuzco and *Hurin* (lower, located in the southeast) Cuzco, and each half into two sectors, or *suyus* (see Figure 5.3). None of these four regions occupies a 90° segment of a circle; the principal rationale for divid-ing the city this way had more to do with the watershed environment and with kinship and hierarchy rather than with considerations of pure geome-try. Lines between *suyus* roughly demarcate the flow of underground water in the Cuzco Valley, which naturally follows a radial plan in this mountain environment. Specifically, *suyus* were intended to serve as an organizing principle to define water rights to the kin-related groups who farmed the wedge-shaped plots of land between the river valleys. The people believed that they received their underground water by right of birth directly from their ancestors, who were believed to reside in the body of mother earth (*Pachamama*) and whom they were required to honor and nurture by making sacrifices to feed her at certain places and times.

Four major roads departed Cuzco, one from each corner of the central square; these served as the dividing lines among the *suyus*. The Inca envi-sioned these lines to extend to the remotest domains of their empire—as far as Quito to the north and central Chile to the south. *Suyus* were ranked, as were the hierarchically organized kin groups who lived within them. The organizing principle of the moiety division, or "halving," was based on whether citizens were located upriver (higher ranking/Hanan) or downriver (lower ranking/Hurin).

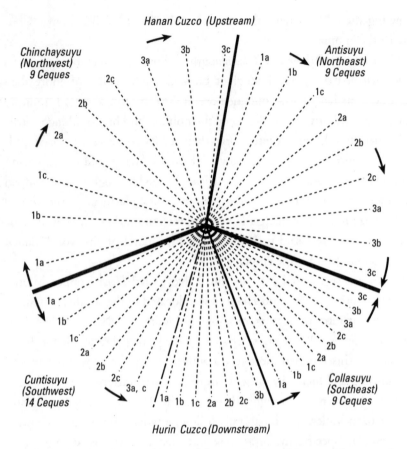

Figure 5.3 This schematic diagram of the *ceque* system of Cuzco exhibits several principles of organization. It illustrates the dualistic principle of halving and then further dividing the city into quarters as well as the vertical splitting into upper and lower topographical sectors. Straight visual lines called *ceques* (dotted on the map) further subdivide Cuzco into pie-like wedges that ideally radiate from the center outward to the corners of the universe.

The hallmark of urban social organization in Cuzco was the *ceque* system. The *ceque* system was a giant cosmographic map, a mnemonic device built into Cuzco's natural and man-made topography, that served to unify Inca ideas about religion, social organization, calendar, hydrology, and astronomy. The Spanish Jesuit chronicler Bernabe Cobo, writing in 1653, has left us in his *History of the New World* a thorough and detailed description of the system, which also may have been adopted as a basic organizing principle in other cities of the empire.[9]

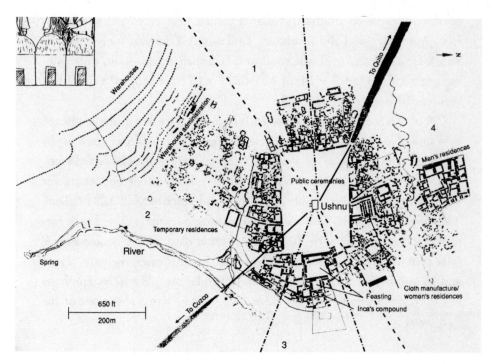

Figure 5.4 Though Old Cuzco was largely destroyed, the plan of Huanuco Pampa, an Inca colony, exhibits many of the same radial features described by the Spanish chroniclers who visited the capital, including a moiety structure divided by a pair of major roads (one leading to Cuzco, the other to the northern capital of Quito); four *suyus* (1) (2) (3) (4), and perhaps even visible *ceque* lines between compounds. Buildings in the royal (Inca) compound are oriented to the direction of sunrise on the day of zenith passage as it would have been observed in Cuzco.

> From the Temple of the Sun as from the center there went out certain lines which the Indians call ceques; they formed four parts corresponding to the four royal roads which went out from Cuzco. On each one of those ceques were arranged in order the guacas and shrines which there were in Cuzco and its district, like stations of holy places, the veneration of which was common to all.[10]

Testifying to the extraordinary power of the organization of the Inca state, Cobo goes on to describe and locate in detail these imaginary radial lines (*ceque* means "ray" or "sight line" in Quechua). He indicates that they were grouped zonally according to their location within each of the four *suyus*. These *ceques* emanated from the Coricancha at the city's center (see Figure 5.4).

According to anthropologist Tom Zuidema's interpretation of the record, Cobo lists nine *ceques* associated with each of three of the *suyus*: the northwest (centered around the district named Chinchaysuyu), the northeast (Antisuyu), and the southeast (Collasuyu). Fourteen more *ceques*, he says, were associated with the southwest (Cuntisuyu) quadrant, thus making a total of 9 + 9 + 9 + 14, or 41 radial lines. I have adopted a simplified version of his map in Figure 5.3.

Cobo's description implies that each *ceque* was traceable in the landscape by a line of *huacas* [*guacas*]. A *huaca* (see Figure 5.5) is a sacred place in the landscape, an opening in the body of *Pachamama* where she can be fed sacrifices. The situation of these *huacas* must have been rather important, for Cobo goes through the trouble of locating and describing all 328 of them. They included temples (natural or man-made); intricately carved rock formations; bends in rivers; fields; springs or other natural wells, called *puquios*; hills; even impermanent objects such as trees. In most cases, the water theme and its association with the agricultural calendar are given heavy emphasis in Cobo's description. Here, for example, is what he writes about one of the 328 *huacas*:

Figure 5.5 An Inca ruler talks to his huacas

The seventh [*huaca* of the 8th *ceque* of the Chinchaysuyu quadrant of the city] was called Sucanca. It was a hill by way of which
the water channel from Chinchero comes. On it, there were two
towers as an indication that when the sun arrived there, they had
to begin to plant the maize. The sacrifice which was made there
was directed to the sun, asking him to arrive there on that hill at
the time which would be appropriate to planting, and they sacrificed to him sheep, clothing, and miniature lambs of gold and
silver.[11]

Note the rather concrete information about both the flow of water and the
sky in this passage and the way information about the environment is
directly tied in with aspects of everyday life.

Each *ceque* was assigned one of a set of three hierarchical groups that represented the social classes ("partialities" and families, Cobo calls them) that
were required to tend to its *huacas*. There were: (a) *ceques* said to be maintained and worshipped by the primary kin of the Inca ruler; (b) those that
were worshipped by his subsidiary kin; and (c) *ceques* tended to by that segment of the population not related to the ruler.

The servants and attendants of these *huacas* came offering the appropriate sacrifices at the proper times, says Cobo. The assignments on the hierarchy of worship rotated sequentially; that is (a), (b), (c); (a) (b), (c), and
so on, as one proceeded from one *ceque* to the next, all the way around the
horizon in a clockwise direction in the northern *suyus* and counterclockwise
in the south, as shown in Figure 5.3 on page 156.

The organization of communal work activities, particularly having to do
with agriculture and irrigation, was also prescribed by the *ceque* map; for
example, representatives of each of forty families drawn from the *suyus* par-
ticipated in the ritual plowing that took place annually in the plaza of
Hurin Cuzco, each delegate plowing a designated portion. Even rules for
the servicing and maintenance of the irrigation canals were specified by the
order of the *ceques*. This idea of physically dividing up a central ceremonial
place as a field for negotiating social relationships remains alive in contemporary Andean communities.[12]

Cobo's description of the *huaca* given above offers our first clear reference
to astronomy, specifically to watching the sun at the horizon. Another

chronicler is even more specific: Felipe Guaman Poma de Ayala tells us that the Inca built observatories with windows "to catch the first and last rays of sun at horizon." They employed these observations to know when to shear the llamas, sheep, and alpacas, and when to sow and harvest crops—he tells us that all these events occurred at different seasons of the year. In other passages in which Guaman Poma describes the duties of office of the state astrologer, he refers quite explicitly to the changing aspect of the sun in the local environment. The sun, he remarks, "sits in his chair one day and rules from that principal degree [of the December solstice]. Then he sits in another chair where he rests and rules from that degree [of the other solstice]." From one seat to the other "he moves each day without resting"— likely a reference to the succession of rising and setting points on the horizon. During the solstices he rests for more than a day in his chair, as we know on those days the motion of the sun from day to day on the horizon becomes imperceptible. Guaman Poma goes on to say that "in the month of August, on the day of St. John the Baptist, he seats himself in another seat; in the first seat of his arrival. . . ."[13]

The historical record clearly helps us to establish both a purpose and a methodology for Inca astronomy. Apparently they had developed a sophisticated system of horizon-based observing, a system that was both practical and functional.

How real is the *ceque* system? Are there still traces of it? Can we map it? Does it appear exactly as Cobo describes it? From Cobo's and other chroniclers' descriptions, investigators have attempted to answer these questions by undertaking a thorough reconnaissance of the Cuzco landscape. Using Cobo's rich descriptions as a guide, they have learned that several *ceques* still can be traced today. Following up on earlier fieldwork by anthropologist Tom Zuidema and me, archaeologist Brian Bauer has undertaken the most careful and thorough attempt at mapping the *ceque* system to date.[14] His method consists of archaeological field surveys coupled with interviews of local inhabitants in order to confirm toponymies that might resemble those mentioned in Cobo's descriptions. In some cases Bauer has been able to locate specific shrines at point locations, while in other instances he has only been able to indicate the general area corresponding to a particular *huaca*. A major conclusion of Bauer's fieldwork is that the *huacas* of given *ceques*, when connected together in the sequences described by Cobo, do not form perfectly straight lines, as a strict reading of Cobo might imply. (A segment of one of Bauer's maps is shown in Figure 5.6.) Some *ceque* lines deviate wildly and even run back on themselves.

So, was the *ceque* system conceived ideally as an organization of straight radial lines that when actually traced over the bumpy landscape became distorted? In Chapter 2 we raised similar questions about our old lunar-based calendar, which in reality is not exactly a series of lunar synodic months, all of equal length corresponding to an observed lunar cycle, though the moon still underlies its origin. These considerations of differences between the *ideal* and the *practical* have a direct impact on our understanding of the exactitude of the Inca astronomical observing system. Employing the revised *ceque* map, Bauer, assisted by astronomer David Dearborn, has attempted to reevaluate some of the earlier connections between the historical record and Inca astronomy.

ASTRONOMICAL ALIGNMENTS IN THE *CEQUE* SYSTEM

One of the most important orientations in Cuzco had to do with the sun's position during the planting season. An anonymous chronicler tells us that the Inca constructed four pillars on a high hill overlooking Cuzco from the west; the inner and outer ones were separated by 200 paces, while the two middle ones were 50 paces apart:

> When the sun passed the first pillar this began the time when they were warned about the planting of vegetables at the highest altitude. When the Sun entered the space between the two pillars in the middle it became the general time to plant in Cuzco, always the month of August. And when the sun stood fitting between the two inner pillars they had another pillar in the middle of the plaza ... called the Ushnu, from which they viewed it. This was the general time to plant in the valleys adjoining Cuzco.[15]

Though there is some doubt about the number of pillars or towers involved, both the anonymous chronicler and Cobo appear to be talking

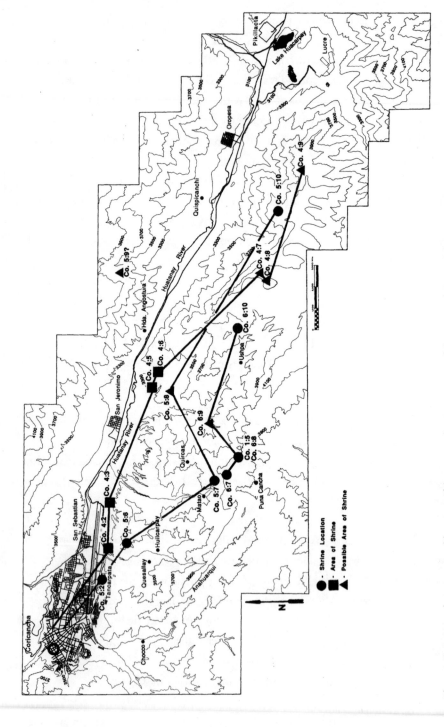

Figure 5.6 Map showing actual location of *huacas* and *ceques* in one of the *suyus* of Cuzco (cf. Figure 5.3, p. 156).

about the same alignment scheme. Also, the anonymous chronicler, like Guaman Poma, mentions August as the key month.

We should not be surprised to read conflicting reports about the number, nature, and location of these marker pillars among the several post-Conquest historians who wrote about them. Like the holy men from Europe who wrote about the Maya, none were known to be well endowed with a knowledge of the sky environment; furthermore, no one of Spanish origin writing at the time of the conquest would be too quick to credit his presumably dull-witted, pagan subjects with a knowledge of matters as esoteric as astronomy and calendar. In the eye of the invader such pursuits would have been seen as totally incongruent with the views they held of the aborigines as inferior.

The hill to the west of Cuzco is the prominent mountain named Cerro Picchu, the summit of which lies two kilometers from the present town square where the *ushnu* was located. Today a cathedral overlooks the *ushnu* site, but antizenith sunset (August 18 in the latitude of Cuzco) still occurs on a precipitous downslope just north of the summit (see photo in Figure 5.7). At the supposed distance from the site of the cathedral, the inner pillars would frame 1° of the horizon (two solar diameters), while the "200 paces" from the first to the last pillar translates to 4°, or eight sun disks, of angular separation. This means about fifteen days would have been required for sunsets to migrate across the zone marked by the pillars, thus providing a reasonable delay interval for planting at different altitudes in the vicinity of Cuzco. Garcilaso also talks about the pillars (which he says he still saw standing in 1560) as being arranged in groups of four; however, he implies that the Inca ruler had erected more than one set of them and that they were used to mark the solstices.

How to interpret these alignments? We know that one of the most important timings in the Inca calendar was the traditional date of the start of the planting season, a day occurring in mid-August when the sun arrived at its position opposite the zenith. Evidently the times for planting in different terraced elevations in the vertical environment of Cuzco were marked out by the day-to-day horizontal course of the sun across the row of pillars. Zuidema proposes that this day could have been incorporated into the *ceque* system via the reversal of an observation taken along a baseline to the rising sun on the day it passed the zenith, the reciprocal of the antizenith date.

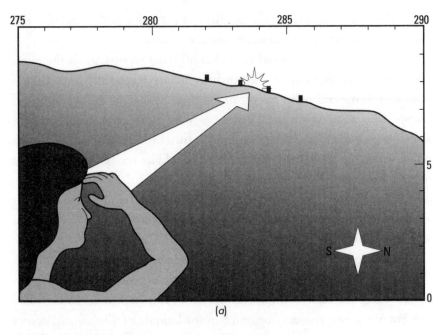

Figure 5.7 (a) Horizon scheme for sighting the movement of sunsets during the mid-August planting season (scale in degrees). (b) Photo shows horizon as it looks today. The antenna in foreground marks the location of the central pillars on Picchu mountain.

The discovery that these important dates coincided with one of the prominent, conveniently visible celestial phenomena in the environment of Cuzco probably led the Inca to regard the act of planting and sun-under-the-world as "going together." At this time of year, the earth mother was opened up. Pachamama was at her peak of fertility and could be penetrated by both man (with his plow) and the sun (with its rays). Viewed in this light, the sight line across the landscape connecting the horizon sun sightings on the days of overhead and underfoot passage becomes a reflection in horizontal space of the vertical-dual organizing principle we talked about earlier that was so common in integrating Andean society. Oddly enough, of all the indigenous calendrical concepts researchers have proposed, the zenith–antizenith alignment has raised the most skepticism;[16] but such doubt may be due in part to the unfamiliarity European observers would have had with such a concept and the way they would have reported it—or indeed neglected to report it. To follow the argument as Zuidema and I have reconstructed it necessitates a descent into the environmental calendar of Cuzco coupled with a trip into the inner confines of the Coricancha to look at the stars as well as the sun.

The association of planting, irrigation, and the Pleiades probably developed when people recognized an approximate equivalence between the period of absence (about 37 days) of the prominent star group from the sky and the time between the end of the harvest and the beginning of the next planting season. This "dead time" likely went uncounted in the annual calendar. There are numerous examples of calendars throughout the world that simply do not keep track of time at the end of the year, usually because no activities take place then. One of the most fascinating examples occurs among the Trobriand Islanders of northern New Guinea. They list only ten months in their year, whose names they correlate principally with horticultural activities. They mark these months via a series of astronomical and biological time checks. One difference between the Inca and the Trobriand calendar is that in the latter the primary event that restarts the year is not the appearance of a star but rather that of a worm! If ever there were a case of humans adapting their own clock to one of nature's most reliable biological rhythms, the relationship between Trobrianders and palolo worms is it. This worm, an inhabitant of Pacific coral reefs, executes a "circa-lunar" rhythm. Once a year for three or four nights, the posterior parts of the worm wriggle dramatically on the surface of the water and let loose their genital products. The spawning marine annelid is seen on the surface of the sea at the southern extremity of the island chain every year following the full moon that falls between October 15 and November 15 (our time). Trobrianders name this important month *Milamala* after the worm, and they cel-

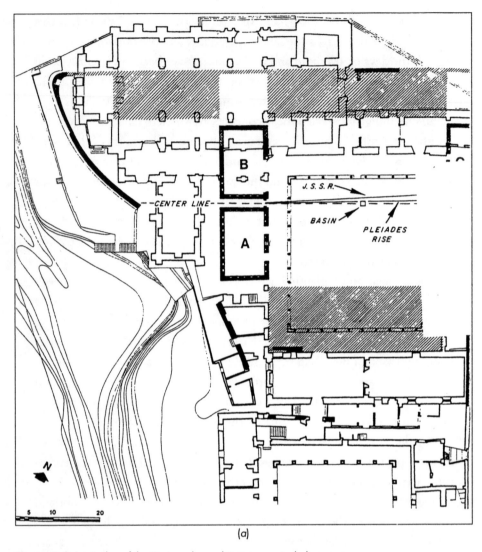

Figure 5.8 (a) A plan of the Coricancha and its astronomical alignments.

ebrate a great festival in its honor to inaugurate the planting season. (They also eat the worm roasted, which they prize as a delicacy!)

The Pleiades trigger the start of the Inca year the same way the appearance of the palolo worm restarts the calendar of the Trobrianders. To understand the rather subtle involvement of the Pleiades in the architectural alignment scheme of the Coricancha, we must return to the focal point of the *ceque* lines, the place in the center of Cuzco where the waters of the city's two rivers comingle. The most important chambers in the Coricancha were

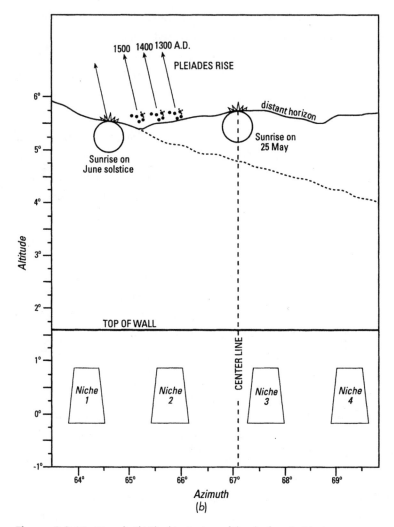

Figure 5.8 (*Continued*). (*b*) The king's view of the sky from inside the Coricancha.

four magnificent halls, which fortunately remain undamaged by many past earthquakes. In Figure 5.8a I have labeled the larger pair of freestanding western structures A and B. Between the two entrance gates on the west side lies a large niche with holes bored at various places along the edge. Curiously, the enclosure wall does not form 90° angles at the corners. Also, the east and west walls are actually antiparallel by about 2°. Yet the corners of each individual building are perfectly rectangular. Why would Inca architects create such an obvious asymmetry? (Recall that we asked the same question about skewed Maya edifices.) Measurements of the alignment of the walls of the Coricancha reveal a solstitial orientation. But then, such a discovery is not so surprising, for the chroniclers tell us that the Inca king worshipped the rising sun there. Moreover, as we shall see, there are already solstitial orientations in the *ceque* system.

If we plot the contour of the eastern horizon in the vicinity of the eastward-directed alignments we can see what the king saw from his niche over the eastern structures over 500 years ago (Figure 5.8b). Sunrise at the June solstice would have occurred about three sun disks to the left (north) of the perpendicular to the west facade (vertical dotted line). This difference, translated into daily solar motion along the horizon, corresponds to a sunrise position about 30 days before or after the June solstice, or to the date pair May 25 and July 20. But our alignment also points close to where the Pleiades rose. (I have plotted their positions for three different epochs in Figure 5.8b.)

What sense can we make of these dates? The first one, which falls one lunar synodic month before the solstice, is in excellent agreement with the timing of the start of the new year according to the chronicler, Cristobal de Molina. He begins his account of the calendar by stating that:

> They [the Inca] began counting the year in the middle of May, with the first day of the moon; which month of the beginning of the year they called Haucay Cusqui, in which they celebrated the ceremonies called Intiraymi, which means "Feast of the Sun."[17]

The "middle of May" in the Julian calendar corresponds to May 25 in the Gregorian. The "first day of the moon" probably meant the day of the moon's first reappearance as a thin crescent in the west after sunset. With such a definition of the "Feast of the Sun," it seems clear that the Inca preferred to schedule activities via a full moon on or around the June solstice.

Our measurements revealed that the direction that the Coricancha faces, both toward sunrise about May 25 and roughly toward the Pleiades, also coincides with the departure direction from the Coricancha of a *ceque* that points close to where the Pleiades rose. Its terminal *huaca* is called by Cobo *Susumarca*, which is the same name as one of the names of the Pleiades. Calculating the morning heliacal rise date of the Pleiades at Cuzco in Inca times, we discovered that this event occurred about one-half lunar month before the June solstice, or about June 8. Thus, if the Inca king sighted the first visible crescent moon falling after May 25, he would have found on the average that the first full moon after that date would fall on the June solstice. In other words, *the first full moon after the first annual predawn rise of the Pleiades always will define that month that includes the June solstice.* Moreover, this moon would be observed on the average not only close to the time of reappearance of the Pleiades but also in the same general area of the sky.

Like the Trobrianders' observation of the palolo worm to restart their seasonal cycle, here, too, was a clever and quite simple set of cross-checked timing devices involving sun, moon, and stars, all of it incorporated into the royal architecture.[18] The king could conveniently and confidently observe these sky phenomena from his golden throne in the Coricancha, and know precisely what was about to unfold in the heavens over Cuzco. He could use the information to prescribe the conduct of human affairs and to validate his connection with nature, the ultimate source of all power.

We can only speculate about the ceremonies attending the festival given over to the worship of the sun on those special days of its cycle when it reached its seasonal high points. Here was the place that divided all of space and time, where high priests convened among the images of their deities and people processed outward to give offerings to their *huacas.* All over the kingdom on the day of the Inti Raymi (June solstice) festival people sacrificed llamas, guinea pigs, and children, and burned textiles. Like the Aztecs and the Maya, drawing to a close a 52-year round of the calendar, they ritually cleansed themselves in cyclic renewal to ensure the health and wealth of both king and empire and to reinforce their participatory role in running the cosmos. There they renewed their mythic connection to their venerated ancestors, children of the gods Inti and Viracocha, who emerged out of a cave to found Cuzco.

THE *CEQUE* SYSTEM
AS A CALENDAR

The *ceque* map was not just a directional scheme that incorporated significant astronomical horizon events. Anthropologist Zuidema has argued that it also was a seasonal calendar with each *huaca* representing one day, and clusters of *ceques* signifying lunar months. The number of *ceque* lines (forty-one of them), doubled, would serve as a count of three sidereal lunar months ($3 \times 27\frac{1}{3} = 82$ days; recall the importance of this period in our discussion of the month of the stars, page 33). And the number of *huacas* in the *ceque* system, 328, is a good approximation to twelve such months. Further, the disappearance period of the Pleiades[19]—the "dead time" in the seasonal calendar when the fields lay fallow—is approximately equal to the 37-day difference between the seasonal year of 365 days and the year of 328 days given by the count of the *huacas*.[20]

The evolution of the Inca calendar is quite complex (see the scheme in Figure 5.9) and many of its details have yet to be worked out, but there are

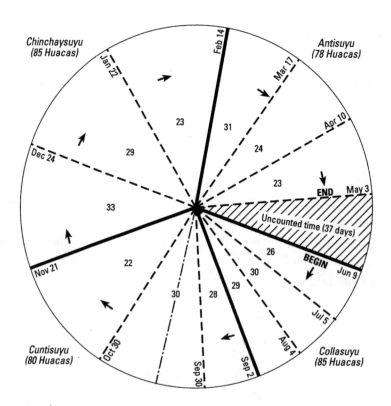

Figure 5.9 The *ceque* system as a calendar: The day count begins with the heliacal rising of the Pleiades (about June 9) and proceeds clockwise; each *huaca* counts for one day.

hints that like the Maya calendar, it may have descended from an orally transmitted counting system based on moon cycles. In the sets of subdivisions of *ceques* and *huacas* given by Cobo's description, Zuidema was able to recognize vague hints of both lunar synodic and sidereal periods. Thus, the first three *ceques* grouped together have twenty-six, the next three thirty, and the next three twenty-nine *huacas* (these average 28.3 or 1.2 days short of three lunar synodic months). Later in the sequence the intervals are further segmented; for example, into 4 4 5 5 5 and 4 4 3 4 4 intervals. Interestingly, sixteen of the forty-one Inca intervals are handsful (5) or multiples of handsful (10, 15) and nine more lie within one unit of these multiples.

Zuidema thinks the Inca adjusted these month periods slightly in order to force them to correspond to significant periods in the agricultural cycle, such as the times of plowing and planting, the appearance of water, and harvesting. We make similar adjustments, but for different reasons. For example, the way we combine Lincoln's and Washington's birthdays into a single Presidents' Day or celebrate Columbus Day on the Monday nearest October 12. As in the Coricancha alignment, there was much concern that certain feast and ritual activities be celebrated during a period commenced by a named full moon. This makes sense when we compare it with the old tradition of beginning the harvest during the month initiated by the appearance of the harvest moon or of opening hunting season with the hunter's moon. The twelve months of the lunar synodic year were divided among the population of Cuzco and vicinity in such a way that each *suyu* was assigned the responsibility to tend to the ceremonies and sacrifices that were performed at the *huacas* that were associated with their particular month of the year in the day-counting scheme. This likely helped promote social integration by giving each family a special role to play at a designated time.

Like the Stonehengers, the Inca also possessed an interest in timing the standstill positions of the sun. Given the difficulty of determining the exact date of the solstice from horizon observations, it would have been reasonable for southern hemisphere astronomers to devise a configuration of multiple alignments to mark the slowing of the sun as it reached its June (winter) limit and its December (summer) limit. Among Cobo's description of the *ceques*, we find two passages that allude to the possible use of just such solstitial pillars.

Chinchaysuyu—*ceque* 6, *huaca* 9:

A hill called Quiangalla that is on the road to Yucay where there were two monuments or pillars that they had for signs and when the sun arrived there it was the beginning of summer.

Cuntisuyu—*ceque* 13, *huaca* 3:

Chinchincalla is a large hill where there were two monuments at which, when the sun arrived, it was time to sow.[21]

How do archaeoastronomers trace out the alignments involved in each of these cases in the field? Figure 5.10 summarizes some of the results. Let us take Cobo's second statement first because it is simpler to analyze. The Chinchincalla twin pillars probably were fashioned to encapsulate the setting sun at the December solstice (which is the time to finish planting). The most likely observation point would have been the Coricancha. The line from the Coricancha to Chinchincalla can be drawn fairly accurately across the land-scape from Cobo's description of the *huacas* on three closely spaced *ceques* in

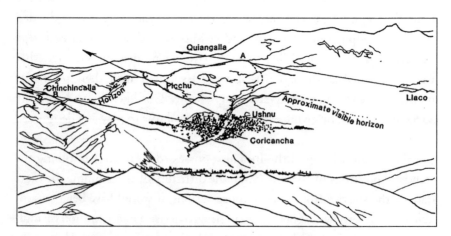

Figure 5.10 Astronomical alignments in the environment of Cuzco. View of the valley of Cuzco from the south showing three horizon alignments constructed in the landscape. Two sets of pillars (A and B), viewed from separate stations in the city, marked the sun at its June and December limits. Four pillars at C marked the sunset point on the day it passed the nadir (antizenith).

this region of Cuntisuyu quadrant. The line coincides with the direction of the setting sun at the December solstice. We can mark out Chinchincalla's northern limit by our relatively precise determination (from local informants) of the location of Pantanayoc, a *huaca* on the adjacent *ceque* 14 to the north, which Cobo describes thus: "It is a large hill parted in the middle that divides the roads of Chincha and Condesuyu."[22] Ravaypampa, the next *huaca* of *ceque* 14 of Cuntisuyu in the direction of Cuzco, is described as a "terrace on the slope of the hill of Chinchincalla," while the *huaca* Puquincancha limits our view on the next *ceque* (number 12) to the south. If we draw a straight line from Pantanayoc to the Temple of the Sun, it passes over a steep slope that is likely the northern flank of the Chinchincalla hill. The December solstice pillars must have been located somewhere higher up on that same slope, where *ceque* 13 passes between *huacas* we are able to locate on *ceques* 12 and 14. Though we found no archaeological remains to mark the spot when we made our measurements in the late 1970s, more recently Brian Bauer and David Dearborn have identified a possible location for the Chinchincalla pillars on the western slope of Killke Hill, a few hundred meters north of the area we explored.[23] There they found a 5-meter-long stone terrace on which bits of Inca pottery had been scattered about some heavily looted burials. The alignment from Coricancha to this point corresponds to an early December sunrise date. But the solstice limit would strike it precisely if the backsight were moved to Intipata, another "Temple of the Sun" immediately in front of the Coricancha. Bauer and Dearborn also have proposed other possible locations for these pillars.

What about Quiangalla, Cobo's other reference to horizon pillars? This hill can be fairly accurately pinpointed on Cobo's *ceque* 6 of Chinchaysuyu, on which it is the ninth *huaca* counting outward from the Temple of the Sun. But once again, neither the Coricancha nor any other monument near it could have served as the observing station, because not all parts of Quiangalla are visible from the center of the city. Tracking the solstice sunset direction from Quiangalla all the way back to Cuzco, we found that it passes a little over a mile north of the center of the city. If long-distance June solstice sunset observations were marked by a pair of pillars on Quiangalla, the most likely observation point from which to observe the event would have been 3 miles to the southwest in the vicinity of a *huaca* named Chuquimarca. Cobo describes Chuquimarca as a temple of the sun where the Inca king spent his time celebrating the June solstice rituals—the ones that marked the pivotal darkest day of the year in the southern hemisphere, when the sun reached its northerly standstill. Cobo says that Chuquimarca was on or near a mountain named Manturcalla. Although neither name is used anymore, by locating the surrounding *huacas* and using colonial documents that mention Manturcalla in the context of other place names there, Zuidema and I found the most logical candidate for the temple in a complex today called Lacco. It consists of a sculptured rock with caves and ruins of buildings, terraces, stairways, and an irrigation system carved in stone. Given Lacco as backsight for the alignment, we traced the June solstice sunset point west-ward from it. On the way to the Coricancha this sight line crosses the *ceque* on which Quiangalla lies. This gives us the foresight and therefore the sup-posed location of the Quiangalla pillars. Though the pillars have long since been dismantled, an archaeological probing for some remains might yet prove fruitful. To date Bauer's intensive survey has failed to turn up any remains in this area, though he has suggested locations nearby that may be tied to solstitial foresights.[24]

So much for solstice watching. There is little evidence that the Inca astronomers marked out the equinox in the Cuzco environment, though a few chroniclers do refer to it in the context of another system of landscape timekeeping in which they made use of gnomon. Pedro Cieza de Leon men-tions the observation place from which the Inca observed sunset over the hill of Carmenga as a location "where the Inca erected a flat plaza of mor-tar and stone to make solar observations." Is this the *ushnu* of the anony-mous chronicler? He fails to elaborate, though he seems to be talking about a place near the Hanan-Haucaypata, the present Plaza de Armas. Most investigators agree that in this instance the chroniclers probably had misun-derstood horizon markers for shadow-casting devices.[25]

Although these examples of insights into Inca astronomy are complex, nonetheless they give an idea of how archaeoastronomers need to combine historical evidence, astronomical knowledge, and architectural and topo-graphic data in order to acquire information on unwritten Inca astronomy. As a result we discover that at least three astronomical sighting directions emanated from three different points in the Cuzco area. First, four pillars were situated on nearby Cerro Picchu to mark the most crucial point in the start of the planting season; these were centered about the place where the sun sets on the day of passage through the antizenith, for obvious agricultural reasons. Second, Inca timekeepers built a pair of pillars to mark the June solstice sunset point as viewed from Lacco, a complex of rock carvings on a hill north of Cuzco. And finally, they erected a pair of pillars to mark the December solstices as seen from Coricancha, the center of the *ceque* system.

If we feel uncomfortable accepting a system of alignments with three different centers of observation, perhaps we should reflect on Jacquetta Hawkes's statement (in Chapter 3) about "getting the Stonehenge we desire." Like Inigo Jones's or John Aubrey's reconstructions of Stonehenge, Garcilaso's and other chroniclers' descriptions of Cuzco need to be viewed through a Renaissance eye that perceived architecture as symmetrical. Moreover, given the low level of cosmological thought the chroniclers believed the Inca to possess, their record is likely to be quite incomplete. Indeed, when it came to scientific matters, those learned men seem to have turned a deaf ear to their informants.

Like Britain's Stonehenge and Uxmal's House of the Governor, which also were built with the sky in mind, the *ceque* system had many overlapping purposes. We can think of it as a mnemonic scheme that organized space, time, and social hierarchy in Inca society. At the same time hydrological principles and kinship also contributed to the structure of Cuzco's urban plan. Further complicating the problem is the issue of straightness. Where does this argument about straight *ceques* come from? That it is not a purely Euclidean idea rooted in the minds of modern Western researchers but instead is a genuine indigenous concept seems clear, as clear as the fact that our monthly calendar is ideally based on the phases of the moon, even though not a single one of the twelve months that make up the year has a length equal to the actual, measured phase cycle. In Quechua the word *ceque* (Spanish *raya*), the name given by Cobo's informants, means "straight" (ahead). Differences of opinion among investigators today seem to revolve around how close to straight the lines really were in practice. The work goes on, and the answer to the question How straight were the *ceques?* will have definite consequences, one way or the other, for future theories about astronomical alignments.

I think what stands out most about Inca skywatching is the way they used the natural order they perceived in the landscape as a means of structuring their social order. The *ceque* system functioned like clockwork only when each particular class performed its assigned function in the proper place along its *ceque* line at the correct time of year. Thus, the people of Hanan-Cuzco were in charge of the June solstice observations and offerings, while those in Hurin-Cuzco took care of the December solstice.[26] In this confrontation between nature and culture, in a harsh, difficult, and changeable agricultural environment, the *ceque* system served to map out and prescribe proper human action based upon residence and kinship in a radial, subdivided geographic framework. It was, like Jorge Luis Borges's fictive description of cartography gone awry, a virtual map of itself:

> In that Empire, the craft of Cartography attained such Perfection that the Map of a Single province covered the space of an entire City, and the Map of the Empire itself an entire Province. In the course of Time, these Extensive maps were found somehow wanting, and so the College of Cartographers evolved a Map of the Empire that was of the same Scale as the Empire and that coincided with it point for point. Less attentive to the study of Cartography, succeeding Generations came to judge a map of such Magnitude cumbersome, and, not without Irreverence, they abandoned it to the Rigors of sun and Rain. In the western Deserts, tattered Fragments of the Map are still to be found, sheltering an occasional Beast or beggar; in the whole Nation, no other relic is left of the Discipline of Geography.[27]

CHAPTER SIX

THE WEST
VS. THE REST?

What is the origin of the peculiarly scientific basis of our own high civilization? In our present generation we many stand on the shoulders of giants and examine in considerable detail the history of science in China, the complexities of Babylonian mathematics and astronomy, the machinations of the keepers of the Mayan calendar, and the scientific fumblings of the ancient Egyptians. Now that we have some feeling for what was possible (and what not) for these peoples, we can see clearly that Western culture must somewhere have taken a different turn that made the scientific tradition much more productive than in all these other cases.

—Derek de Solla Price, *Science Since Babylon,* 1975

At the outset I spoke of a barrier that always seems to stand in the way of our comprehension of other cultures, specifically their astronomies. That barrier is both wide and deep, for the development of Western science teaches us how to comprehend the natural world in a certain way that has been dominant in our society for more than 500 years, particularly so in the present age. Being centered in our own way of thinking leads us to presuppose that there is, after all, only one real world and everyone ought to comprehend it the way we do. We would do well to look briefly at the underpinnings of the scientific astronomy developed in Western Europe so that we can place our sketches of Bronze Age, Mesoamerican, and Andean astronomies in proper contrast.

Expressed by mathematical laws that we believe describe how things work—laws we have faith we can discover and map out—today's astronomer seeks the ultimate causes of the events we all witness taking place in a vast, space-bound, orbit-filled universe. Where did we acquire this way of knowing the universe, and how were these ideas communicated to us? Modern experimental science was born of the European Renaissance, a time when skeptics questioned the existence of a cosmos manipulated strictly by the hand and mind of God. But we can trace science's roots back even further in time to the ancient Greeks, who gave us tools for formulating a new concept of the universe. They offered their descendants the gifts of logic and reason in the form of geometry—our spatial way of understanding things. In turn the Greeks had received gifts of their own, especially from the Babylonians, whose arithmetical way of thinking was passed on across the Aegean.

Historian of science Derek Price suggests that our unique way of comprehending the natural world began with the marriage of the Babylonian quantitative approach to planetary motion and the Greek tendency to form qualitative and pictorial models rooted in a system of formal logic. Babylonian expert Asger Aaboe goes further. He isolates the Babylonian astronomical system as the first true *scientific* astronomy because it could be formulated as "a mathematical description of celestial phenomena capable of yielding numerical predictions that can be tested against observations."[1] The Greek achievement lay in retaining the picture of planets that move in perfect circular orbs as the guiding concept, then fitting all of the observed planetary motions to that pictorial mode; they were the first designers of machinelike models of the universe. Before dissecting Greek model-making, we should look at what the Babylonians contributed to the Greek legacy.

GIFTS FROM THE EAST

Of all the cultures that made our astronomy what it is today, the Babylonians have left us perhaps the clearest record of their donation—all of it in a once-alien writing form that only in the twentieth century has been fully deciphered.

Historians could have anticipated that the Babylonians would invent writing on clay tablets—that they should hammer curious juxtapositions of triangular impressions we call *cuneiform* writing (*cunei* means wedge-shaped) onto wet, viscous earth compounds, allowing them to harden into eternal permanence. Clay tablets were the logical medium to carry astronomical as well as other information—legal documents, licenses, tax bills, and so on. Greasy clays were abundant all over the land between the Tigris and Euphrates rivers, especially after the annual flooding occurred. As the mud dried and caked, it created a natural medium on which to place visible symbols that could help people record things past. Babylonian cuneiform tablets are still capable of telling us today what these people thought about more than two thousand years ago. Their laws, their stories of creation, their dynastic histories, even the passage of time are preserved in cuneiform. Just as they once were positioned in city squares and ancient libraries, so today the clay tablets reside in museum cases to be read by all the people.

According to art historian Denise Schmandt-Besserat, neither mathematician nor astronomer invented cuneiform.[2] It resulted from an expanding market economy. She believes the innovator probably was a wise farmer or trader. In bringing jars of oil to trade for sacks of barley and dates, there is a problem. How to remember and record who owes what to whom? This is especially difficult if large quantities are shipped long distances. Researching the world's first bills of lading, which date from 3350 B.C. and come from excavations in Uruk and other cities in Iraq, Iran, and Syria, Schmandt-Besserat helped crack the economic mnemonic code. These account books consisted of clay envelopes that held collections of small clay tokens of various shapes. Each configuration represented a particular kind of trade item; for example, cones might represent grain, disks could be animals, and a sphere with a hole in it a unit of land measure. Once a transaction was agreed upon, the envelope was sealed and its contents embossed on its outside in two-dimensional replicas of the shapes that lay within. After generations of such use somebody conceived of the efficient idea of allowing the record on the outside of the envelope to stand alone as a statement of what lay inside. The bulky envelope was flattened to two dimensions and, voilà, the first writing tablet was created!

The astronomical cuneiform tablet in Figure 6.1 is an early example of Middle Eastern writing. Dug out of the ruins of the library of King Ashurbanipal of Nineveh by British archaeologist Sir Austin Layard in 1850, the repetitive quality of the tablet's text quickly enabled epigraphers Sir Henry

Figure 6.1 Babylonian cuneiform tablet. Tablet 63 of the series Enuma Anu Enlil, the so-called Venus Tablet of Ammizaduga, sixteenth century B.C.

Rawlinson and George Smith to decipher its basic content. Like most antiquarians of the time, they were searching for the Tower of Babel, the birthplace of Abraham and other biblical artifacts. What they discovered, to their surprise, had to do with neither oil nor grain but something far more esoteric. Like many others of its kind, the tablet tells us about the types of astronomical records Babylonian astronomers kept and how they used their observations to make celestial predictions. The scheme for marking time as stated in the tablet should be familiar to us because we have encountered it before—in our study of the Maya. In each written statement, the planet Venus is said to disappear on a particular day of a given month and to return on another. The observations refer to the times when Venus disappears in the light of the sun as evening star and when it reappears as morning star, and the converse—none other than the four heliacal events we discussed in Chapter 2. Between these stations in time the astronomers computed intervals. The Babylonians also marked stations in space (degrees of longitude and latitude) as well as time. This is indicated in other quotations in which, for example, Mars is said to stand in one position on the zodiac and then in another. The interval in between gives the degrees of separation. Also like the Maya, Babylonian astronomers recorded (in another tablet) a series of lunar eclipses from which they calculated the difference between the times of occurrence of similar successive events. It reads in part:

> Venus disappeared in the east the month of Adar 25th, in the eighth year of Ammizaduga.
>
> Venus disappeared in the west the month of Adar 11th; period of absence four days; rose in the east the month of Adar 15th.
>
> Venus disappeared in the east the month of Arahsamna 10th; period of absence two months and six days; rose in the west the month of Tebit 16th.
>
> Venus disappeared in the west the month of Ulul 26th; period of absence eleven days; rose in the east, 2nd month of Ulul 7th . . .[3]

Each statement is followed by an omen indicating heavy rains, hostile attacks, abundant crops, and so on.

In all of these examples, the Babylonian way of forecasting is based on a simple, familiar arithmetical sequence:

$$\text{PLACE} + \text{SPATIAL INTERVAL} = \text{FUTURE SPACE}$$

or

$$\text{TIME} + \text{TEMPORAL INTERVAL} = \text{FUTURE TIME}$$

The first formula charts the future place in the sky where an event ought to be observed, the second (exactly like the Maya) the future time when an event is anticipated. Modern historians of the ancient Near East have devoted considerable attention to the problem of dynastic chronology, and

Tablet 63 of the series Enuma Anu Enlil, also called the Venus Tablet of King Ammizaduga, with its precise and detailed astronomical record, has emerged as a seminal text. The chronology question is important to the historian, not because of anything of note Ammizaduga accomplished but because his predecessor, Hammurabi, most famous for having devised the first code of laws, was a member of the same dynasty.

King Ammizaduga's tablet conceals hard evidence in a chronological detective story. From lists of Assyrian kings and other texts, we know that Hammurabi reigned some time between 1900 and 1680 B.C. and that he preceded Ammizaduga by 146 years. Here is the problem for historians who are interested in the tablet: What hypothetical period for the reign of Ammizaduga is compatible with the details of the astronomical record given in the Venus Tablet? The answer to the question What time period of observation goes with a given ruler? brings together historical data about the relative lengths and sequences of reigns of kings with the absolute data of astronomy. Using simple software packages, modern archaeoastronomers can back-calculate to within a day the times of full moon 3,000 years ago and to within two or three days' observational accuracy the dates of appearance and disappearance of Venus, assuming good weather. While a unique set of real astronomical dates separated by intervals of fifty-six and sixty-four years turns out to match Venus appearances and lunar phases given in the observational record, only one set, the one that best matches history, will yield the truth. Which is it? Today, after almost a half century of deliberation, the jury is still out, but investigators have zeroed in on 1581 B.C. as the most likely choice for the start of the Venus run of Tablet 63 (alternative choices are 1701, 1545, and 1637 B.C.).[4]

King Ammizaduga's tablet contains information that seems esoteric to us, but to a second millennium B.C. courtly priest it is filled with practical knowledge. Would the astrologers who used the Venus events predicted on the clay tablet to cast omens ever have imagined that four millennia later others would use their statements primarily to figure out *when* they wrote them rather than *why?* In the present context we are less concerned with the use of the tablet to set historical chronology and more with the observations themselves, in particular with what they tell us about the astronomy of the people who performed and recorded them. And we are surprised by their resemblance to the way sky information was processed and recorded in the New World Maya documents.

Table 6.1 lists the fourfold Venus intervals derived from the Venus Tablet of King Ammizaduga. I have placed comparable numbers from the Maya Venus Table in the Dresden Codex alongside the actual observed intervals for comparison. The results are strikingly alike! Notice that in the Babylonian case, the 7-day disappearance, like the 8 days the Maya settled on, is pretty close to the real mark, but in both cases the assigned 90 days is impossibly long compared with the observed 50, or at most 75, days of actual absence of the planet from the sky. Recall that the Maya tabulated visibility periods of 236 and 250 days to fit Venus's motion into the unfamiliar rhythmic time frame marked by the phases of the moon. Did the Babylonians harbor a similar idea? At first sight apparently not, because they chose equal intervals of 245 days, which do not jibe with the lunar synodic month. But if we link one of the 245-day appearances in Table 6.1 with the 7-day absence, we get 252 days, which is just one day in excess of 8½ months marked by the lunar phases. Also, the too-long 587-day total comes a little closer than the observable average 584-day Venus cycle to a whole number of lunar months. (It is a little over 3 days short of 20 moons.) We can think of these fabricated intervals in the same way we regarded the neat little prescription Maya astronomers devised to keep Venus appearances

TABLE 6.1. INTERVALS (IN DAYS) USED IN SIMPLE MODELS TO PREDICT FUTURE APPEARANCES OF THE PLANET VENUS

Phenomena	Babylonian Model (Ammizaduga Tablet)	Maya Model (Dresden Codex)	Actual Observable Average
	Days		
Morning Star (visible in east)	245 (8 mos., 5 days)	236	(263)
Disappearance Period (eastern disappearance to western appearance)	90 (3 mos.)	90	(50)
Evening Star (visible in west)	245 (8 mos., 5 days)	250	(263)
Disappearance Period (western disappearance to eastern appearance)	7	8	8 (0–20)
Total Cycle	587 (19 mos., 17 days)	584	584 (583.92)

and disappearances in tune with the moon phases (page 124–125). You could count out any small aberrations on your fingers.

I find it remarkable that two cultures on opposite sides of the globe chose to structure Venus events in an almost identical way. And I do not think this coincidence has anything to do with cultural contact. I think both the Babylonians and the Maya did it because cycles of time manifest in the heavens, though they be many in number, are plainly visible to everyone the world over. That moon cycles would be used by different observers as the fabric on which to embroider patterns of the course of the planet Venus should, after a bit of reflection, come as no surprise to anyone who watches the stars.

The goal of the Babylonian priest who invented the 587-day Venus formula, like that of his counterpart who followed the course of Venus in the skies over Yucatán, seems to have been to establish an orderly and unified method of making future predictions based on past observations. Neither could have escaped wondering why they sometimes needed to deduct or add in days in the Venus count to keep their canon of predictions on target with the heliacally rising and setting planet. To confine conjecture, meditation, and speculation about the heavens only to the present and to ourselves would be too narrow-minded in my view. Still, the record of numbers on both clay and parchment is mute when it comes to the question of what went on in the minds of ancient astronomers, and so I shall say no more on this issue.

Modern scientific predictions work in the same way as those of the ancient Babylonians, or, for that matter, the Maya. We devise somewhat more complex mathematical expressions and we call them the laws of nature. We have an abiding faith that our mathematical statements will generate predictions about the future once the proper time cycle is inserted into them. When the predicted event comes due we carefully watch the sky, record what we see, and compare it with what the formula says should have happened. Then we alter and modify our equations, progressively adding terms, inserting factors that make for more accurate predictions that can be rendered valid in the world of experience, a world in which we seek to detect even more subtle patterns. When one model becomes too complex we finally address the difficult issue of devising and substituting a simpler one in its place.

"When Venus stands high, pleasure of copulation. When Venus stands in her place, upraising of the hostile forces...."[5] So reads a Babylonian omen. As much as we might laud the Babylonians for their careful observations and their astronomical predictive skills, we must never lose sight of the fact that for them, just as for the Maya, the underlying motive for seeking intricate patterns in the heavens was astrology. A frustrated seventh-century-B.C. skywatcher in the royal court lamented:

> The king has given me the order: Watch and tell me whatever occurs! So I am reporting to the king whatever seems to be propitious and well-portending [and] beneficial for the king, my lord [to know].... Should the king ask, "is there anything about that sign?" [I answer], "Since it [the planet Mars] has set, there is nothing...." Should the lord of kings say, "Why [did] the first day of the month [pass without] your writing me either favorable or unfavorable [omens]?" [I answer], "Scholarship cannot be discussed [heard] in the market place!" Would that the lord of kings might summon me into his presence on a day of his choosing so that I could tell my definite opinion to the king my lord![6]

There is no getting away from it: In ancient times, belief in celestial deities wove our destiny so tightly, so intimately, that one could not avoid a preoccupation with their wanderings. The entire cosmos was an expression of wills imposed by animate anthropomorphized forces that made up the state, just like their kingdom here on earth. Every object in the sky was alive with a personality of its own, ready to unleash its power for good or evil on mortals below. That the tides, the wind, and the rain could be predicted by watching celestial events seems reasonable enough, but the health and wealth of kings and peasants? Hardly—at least for us!

Like the Maya, the Inca, and the architects of Stonehenge, worshippers of the heavenly abode of these deities would appeal to them by performing certain rites. The language comprising the dialog between mortal and transcendent consisted of offerings and incantations; the implements of communication were charm and amulet rather than compass and telescope. These people felt closely connected with what was going on around them.

They experienced nature's forces directly: earthquakes and floods, miscarriages and deformities at birth, eclipses and rainbows. They never imagined such phenomena merely as detached events in a universe devoid of meaning. All things happened for a reason—to warn them and to convey a message, either good or bad, that would guide their future. And some phenomena occurred with a more predictable regularity. For the farmer, a moonrise could tell what to anticipate in the forthcoming crop cycle. For the king, an appearance of Jupiter might signal what would take place in future encounters with the people who lived to the east, those over whom he sought to extend his dominion.

THE GREEK LEGACY

We revere the Greek philosophers for teaching us how to theorize, how to speculate upon our existence, how to question with skepticism. Today's astronomers are also indebted to the Greeks for their capacity to build models and to conjure up particular representations of the real world by using a very special form of reasoning: logic based in a geometrical conception of space. How was this agenda acquired, and how was it bonded to the Babylonian arithmetical contributions? Surprisingly, we draw a lesson from the Inca. The roots of all those old earth-centered models of the solar system are to be found not in the abstract world of Greek philosophy or cosmology but instead in their very concrete realm of city planning. In the classical world, city and cosmos were one and the same.

If you look at the way the Greeks planned Athens, you begin to find similarities with the way they believed their gods designed the universe. The Agora—literally the hearth of the city—lies at its core. There resides the source of all the political and economic forces that drive the engine of Greek society. Like the earth within the universe, the Agora is the place in society where all people convened and interacted. On the urban periphery lived the different classes who composed the whole of Greek society—first merchants and shopkeepers; and beyond that ordinary citizens, the freemen who made up the Greek *polis*, each attended by his slaves. (The Greek concept of democracy regarded slaves not as people held in bondage, but rather as commodities necessary to the success of the economy.) Like society, the universe was hierarchically constructed from center to periphery—orb on earth-centered orb. Among intellectuals, the physical universe was as much the subject of discussion and speculation as the structure of the ideal city. A person had license to measure and theorize about the plan of the universe and could speak of it in the same breath as the organization of the city.[7]

Making models for the Greeks meant literally creating mechanisms with interconnected moving parts. They called their models *simulacra*, "simulations" that describe how natural phenomena behave. Greek scholars believed that if they looked beyond the gods they could discover the underlying principles that governed heavenly motion. Ancient Greek models of the universe are as mechanical and machinelike as many of those devised by science today. For example, we say the brain is like a sponge, the nervous system like a computer; and the atom operates like its macroscopic counterpart, the solar system, as a kind of celestial billiard game. The Ionian philosophers who devised such models were pure pragmatists, reared in the free-thinking, highly individualistic social environment of the Greek colonies of Asia Minor and southern Italy in the sixth and fifth centuries B.C. At this particular period in Greek history, power had recently been decentralized from the priestly hierarchy. Now there was no need to dress accounts of the workings of nature in religious garb. It is easy to understand why, in the eyes of the free-thinking architects of modern science in the time of the European Renaissance, these Greek models of how things work seemed so impressive.

Simulacra were based on everyday common sense. For example, Thales of Miletus (624–547 B.C.) postulated that the earth is a flat disk surrounded by water (Figure 6.2a). He reasoned that it must be a disk because that is what the horizon around us seems to indicate. And it must float on water because we recognize groundwater surging up from below in the form of artesian wells. That there is water above us is demonstrated every time rain falls. And as far as traders and explorers could make out, the horizontal dimensions of the earth disk were encompassed by a great ocean. Such a water-bound terrestrial model is both practical and logical. It derives its explanatory power from the common sense of everyday experience.

Anaxagoras's (500–428 B.C.) world model focused more on heaven. He believed in a cylindrically shaped world (we live on its flat-topped surface)

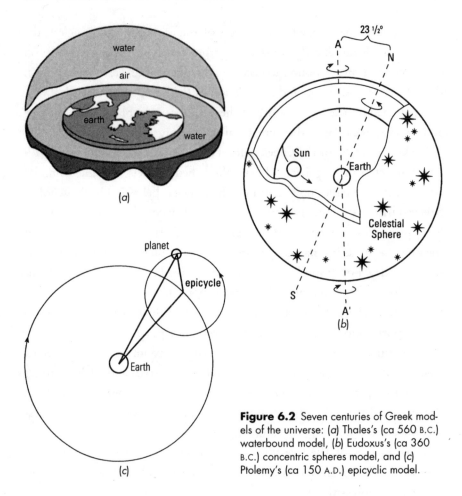

Figure 6.2 Seven centuries of Greek models of the universe: (*a*) Thales's (ca 560 B.C.) waterbound model, (*b*) Eudoxus's (ca 360 B.C.) concentric spheres model, and (*c*) Ptolemy's (ca 150 A.D.) epicyclic model.

that floats freely in space. The stars are appended to a sphere that rotates around the cylinder, carrying them below the cylinder every night, then back into view the next day. The moon shines by the light of the sun. We know this because if we watch its phases we can see for ourselves that the lighted portion of the lunar disk always faces the sun. Eclipses of the moon happen when the earth's shadow falls upon it. You can tell that because we here on earth always seem to be located between the sun and moon whenever an eclipse takes place. Whether you believe Thales or Anaxagoras, the common denominator their way of thinking shares is that simple observations of nature and common sense go together in their models. Note, too, that their explanations are deeply rooted in concepts of geometry and space.

There are no hidden forces and no abstract qualities in the Ionian models of the cosmos. Everything makes sense to anyone who interacts with the natural environment on a daily basis. You get a picture of the world exactly as you see it. In stark contrast to the Ionian philosophers in the colonies, their fifth-century-B.C. counterparts, the Socratic philosophers on the mainland, contemplated nature as a purely intellectual set of interrelated phenomena that operated on certain underlying principles in which they, too, acquired an uncompromising faith. *Underlying* is the key word here, for the Greek thinkers of the Classic period, like our contemporary scientists, believed that what we see on the surface of things is only an imperfect approximation of a deeper, hidden truth that exists in the form of unvarying principles that remain fixed in an ideal world—a world that exists only on a mental plane.

This idealism modern science, too, has inherited. For example, when, in the early eighteenth century, Isaac Newton explained that a cart rolling over a flat surface ought to maintain a constant velocity forever, never coming to rest, he was expressing the idea that in the real friction-free universe hiding beneath the imperfect one we experience, there would be nothing to interfere with a body's movement. In such a universe the natural state of things ought to be constant and eternal motion along a straight line, rather than the ultimate state of rest to which all things seem to come in the world of everyday experience. Newton reasoned that the microscopic bumps on the roadways of our world prevent us from realizing the more fundamental, simpler, underlying state of the universe—a universe without friction. Imperfections such as friction deceive the senses by masking nature's true secrets. This is a way of viewing the natural world that Newton acquired from his Greek forebears.

We have seen that when we shift focus from our Babylonian to our Greek roots we acquire one quality of modern astronomy with which we have become quite comfortable. Imagining celestial bodies as spheres moving about a void, or appended to spherical shells rotating about a common center—this was a characteristically Greek way of expressing how the universe behaves. This space-based geometrical way of perceiving the cosmos lives on today in modern maps of the solar system, the Milky Way Galaxy, even in the Big Bang origin of the universe. It is the source of our concept of space travel, and it also extends in the opposite direction to the microcosm of subatomic structure. This Greek geometric conception of space is based on points and lines, solids and planes.[8] To reduce to a minimum the sum total of observed motions on the sky dome and to perfect a working model of these motions consisting of circles and spheres inscribed into three-dimensional space—this was the ultimate challenge of Greek astronomy from the Classic period forward.

The observed celestial motions that preoccupied Greek model makers were the same as those that challenged Babylonian astronomers, except that the latter, as we have seen, chose to account for them in a primarily temporal rather than a spatial mode of expression. What motions did they observe? We can recall them from Chapter 2. First, the relatively rapid east-to-west motion of every object in the sky in 24 hours. Second, the much slower west-to-east motion of the moon, the sun, and the planets reflected against the background of constellations, each in its own period. And finally, for the planets only, the periodic short-term reversal of this second type of motion during which the planet slows to a halt, turns westward for a time, stops again, and then resumes its normal eastward course among the stars. For the Greeks this so-called retrograde motion, which results in a curious loop for each planet if plotted out against the stars, seemed most beguiling of all. Even by the time of the scientific Renaissance in the sixteenth century, it continued to defy great thinkers like Copernicus and Kepler.

Eudoxus of Cnidus (408–355 B.C.) was one philosopher who sought to meet the challenge of reconciling these motions, a challenge that was said to have been issued by no less thoughtful a philosopher than Plato. First, to give a theoretical explanation of planetary motion, Eudoxus devised a model (Figure 6.2b) that consisted of a set of concentric spheres. Like Aristotle, he believed these spheres were made out of a special crystalline substance, a fifth element confined to the heavens called the "quintessence" (the other four elements were earth, water, air, and fire). He assigned one sphere to each of the seven moving bodies: the sun, the moon, and the five naked-eye planets. Each sphere rotated at such a rate and had its axis inclined at such an angle so that when set into motion the model "saved the phenomena" (for simplicity I have indicated only two of the major axes in Figure 6.2b); that is, it geometrically reproduced the observed motion of each planet appended to it. The outermost sphere, to which all the others were attached, contained the stars. It turned about the earth-fixed center of the universe in 24 hours. While this model did a pretty fair job of returning everything in the sky to approximately the correct place at the correct time, it failed to account for the variable rate of movement of the sun over the course of the year and, worse still, it did not explain why a planet's given retrograde loop differed ever so slightly from the previous one. Only by making his crystalline sphere model more complex, namely by adding additional spheres, could Eudoxus's followers in succeeding generations accurately account for the subtleties of what once were perceived to be rather simple movements.

Awash in a tidal wave of fresh data gleaned from more careful sky observations, the problem of charting planetary motions had become very complex. At one point in the modification of Eudoxus's model, Mars (a particularly aberrant planet) because of its retrograde motion required a total of seven spheres to "save its appearances." Ingenious as it was, Eudoxus's model caved in under the sheer weight of its complexity. As one Greek philosopher remarked, no gods, whether they meddled directly or indirectly in human affairs, could have created a universe so complicated as this one!

Another imaginative cosmic *simulacrum* came out of Alexandria in the second century A.D. Devised by astronomer Claudius Ptolemy, it retained the earth at the center of a spherical universe, but handled retrograde motion in a more elegant way by imagining each planet to orbit on an epicycle, the center of which moved on a bigger orbit, called a deferent (Figure 6.2c). Every time a planet, say Mars, reached the inner part of its epicycle and aligned with the earth, it appeared to execute a brief backward loop—retrograde motion. The effect can be compared to watching a piece of chewing gum stuck to the rim of a moving bicycle wheel. As the rider passes by, the viewer fixes his or her eyes on the path of the gum. When it comes in contact with the ground and the wheel rolls over it, the wad momentarily seems to reverse its direction; then it pitches forward again and passes over the top of the wheel.

Lest we become lost in the Aegean world of ethereal model-making, we once again need to remind ourselves that the ends served by "saving the appearances" were no different from those of Babylonian arithmetical astronomy—they were astrological. The state demanded from those who watched the sky a reading of the omens and predictions on which it depended. Ptolemy is as well known for his astrological writings as he is for his tables of astronomical predictions based on his model. Also, we ought to appreciate the way these models consistently cling to the belief that circular orbits can be envisioned and re-envisioned in a spatial framework to give an account of the movement of things in the sky.

I think it would be inappropriate to label Ptolemy's geocentric universe idea as wrongheaded because the earth *really* isn't the center of everything, or because objects can't travel on epicycles at the center of which there is no material body to attract and hold them in place. We must remember that the concept of gravitation, along with the idea that forces can act at a distance along lines interconnecting celestial bodies, were not even imagined until shortly before the time of Newton, who lived fifteen centuries after Ptolemy. Besides, in the Greek worldview a human being was a microcosm at the bottom of a hierarchy of spheres that interacted with one another. People were a part of the universe and not external to it as most of us have grown accustomed to believe today. Once again, we must resist the egocentric habit of transplanting our notions and ideas across the vast sea of time and implanting them in the heads of our predecessors.

When Ptolemy's model proved slightly inaccurate, he improvised it by putting the earth slightly off-center. In the face of more accurate observations, his followers were forced to add epicycles on top of epicycles. Surprisingly, the Ptolemaic geocentric model of the universe withstood the test of observation well into the scientific Renaissance, when the Polish astronomer Nicolaus Copernicus (1473–1543), disbelieving any model that could be so complex, replaced it with the novel heliocentric (sun-centered) theory of the universe in which all the planets traveled around the sun in circular orbits instead of around a fixed earth. Later, Kepler's analysis of even more precise data revealed that the planetary orbits were better described by ellipses, not circles. Here begins the story of modern astronomy one usually finds at the beginning of most astronomy textbooks.

Clearly, Greek scientific skills were grounded in rationalism as well as in the ability to modify a theory in the face of new information. These sky watchers were erecting the universe they desired—one controlled by hidden principles in the hands of gods who had far more important celestial business to tend to than to meddle in human affairs. Many aspects of this kind of universe are still desirable to us. Modern astronomy leans heavily on the data—the observational evidence that either can foster or fragment highly idealized theories of the universe. In the past two centuries the Industrial Revolution has created an explosive development in the technology of instrumentation to further amplify the role of experimentation and precise measurement. None of the new discoveries could have taken place without precision equipment: telescopes and other energy detection devices, clocks, and information processors.

There is irony in the way contemporary astronomy has developed. As our machinery penetrates ever deeper into the microcosm and farther out into the macrocosm, so, too, it functions to form the barrier I have been talking about between ourselves and our ancestors' view of nature. The complex tools we construct remove us from direct contact with the natural environment our predecessors once experienced. As far as our attitude about the astronomy and calendar of other cultures is concerned, the power of our technology and the scientific outlook that accompanies it can lead to a kind of naturalistic narcissism. Unchecked, such an attitude leads to the belief that if we cannot reduce the astronomy of another culture to some form of our own, then it may not be worthy of our attention. I hope that our excursion through three great astronomies of the past, capped off by a short walk through the underpinnings of our own, has contributed to correcting this false impression.

The sky will forever fascinate all those of us who look up at it. Having produced Alexandrian Greece's most famous book of omens from the sky, in addition to its most celebrated astronomical tabulations, Ptolemy once wrote: "in studying the convoluted orbits of the stars my feet do not touch the earth, and seated at the table of Zeus himself, I am nurtured with celestial ambrosia."[9] My guess is that the exotic astronomers of the ancient Andes, Mesoamerica, and megalithic Britain, whose unwritten records we have only begun to explore, would have understood how Ptolemy felt.

THINGS
TO THINK ABOUT

CHAPTER TWO
THE NAKED SKY

1. Draw a diagram like that of Figure 2.1 on page 12 showing how the sky would be depicted as seen by observers at (a) the Pyramids of Giza (latitude 30° north), and (b) Quito, Ecuador, the northern capital of the Inca empire (latitude 0°).

2. Suppose it is March 20 and you are situated in Cuzco (13½° south latitude), ready to watch the sunrise, but instead of viewing the event on a flat horizon of zero elevation, there is a mountain range approximately 3° high in the foreground. Can you anticipate whether to look to the left (north) or to the right (south) of due east to see the first gleam of the sun's disk? Drawing a picture will help. On sunset that day, where will the sun disappear—to the left or to the right of true west over the western horizon, also presumed to be elevated by 3°?

3. Are the constellations universal? Figure A.1 is a tracing made from a modern star chart, except the lines that traditionally connect the stars together have been removed. What constellation patterns, if any, can you recognize?

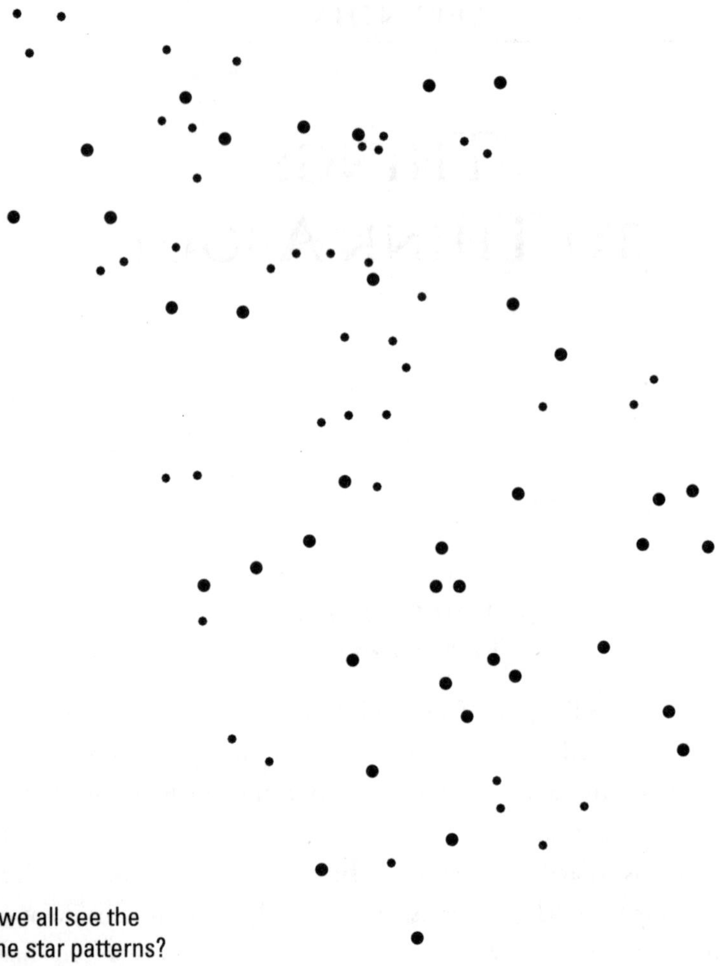

Do we all see the
same star patterns?

Figure A.1

CHAPTER THREE
STANDING STONES AND STARS

1. Are Hawkins's alignments coincidental? Suppose we select the most prominent archaeological features of Stonehenge and set up a hypothetical alignment scheme. Figure 3.3 (page 66) is adapted from Hawkins (*Stonehenge Decoded*, p. 170). How many points has Hawkins selected from/to which he has measured his alignments? Using each point together with all the other remaining points, how many alignments does

he indicate occur in this array of stones? Using each stone as a backsight and then as a foresight, how many alignments are possible? For astronomical targets, Hawkins used the four solar and eight lunar standstills (the moon, which can stretch 5° either side of the ecliptic, would have a pair of standstills associated with each solar standstill). To constitute a "hit" he allowed an accuracy of ±1° for the sun and ±1½° for the moon. Suppose we adopt a liberal ±1½° for both. Assuming no overlap, what are the odds that any one of our possible alignments would hit a target? How many hits would be anticipated from the entire array? Hawkins says he obtained a total of 24 hits (10 for the sun and 14 for the moon) and that the odds that this would happen by chance are millions to one. Do you think this statement is valid?

2. Paleolithic calendar or knife sharpener? The carved piece of bone shown in Figure A.2, part of an eagle's wing, was recovered in a cave in central France nearly one hundred years ago. It has been dated by archaeologists to approximately 30,000 B.C. The comma-shaped marks were gouged out with a pointed tool. (The user probably grasped the bone in one hand and the tool in the other.) Gash marks on the edge were presumably made by drawing another tool over it. Alexander Marshack of Harvard's Peabody Museum has argued that the sequence of points is a notational record of the lunar phase cycle. If he is correct, that would make it by far the earliest record of its kind in human history. Suppose, following Marshack, that each mark stands for a day and that the sequence proceeds as indicated by the line in a back-and-forth sweeping notation, the carver periodically turning the bone every time he or she came to a significant change in the lunar phase cycle, so that the time line would then run in the opposite direction. Look at the sketch and think about what might constitute significant changes in the lunar cycle. Suppose, for example, we define significant changes to occur at new and full moon, which would demarcate the month into its waxing and waning phases. Suppose further that no count takes place a day or two around new moon when the moon is not visible. By counting groups of markings can you see a pattern reminiscent of the moon's phase cycle?

3. Suppose you want to set up an observatory somewhere in Mesoamerica to chart the daily motion of the sun at horizon. You select a pair of sticks, a foresight and a backsight, and position them 100 yards apart.

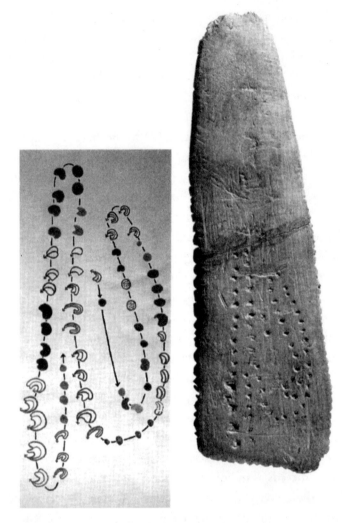

Figure A.2 Carved Paleolithic bone. Was it a calendar or just a simple tool sharpener?

Sinking the foresight into the ground in a vertical position, you then line up the backsight so that the setting sun is positioned over both markers. Then you drive a stake in the ground at the point where the backsight touches the ground. This marks the sunset alignment on day I. One day later you come back to the same place and set up the backsight again. This time the sun has shifted, and as a result the stake marking the spot will fall a short distance to the north or south of the first position (depending on what time of year it is). Table A.I gives the amount of

TABLE A.1. CONVERSION OF LINEAR TO ANGULAR AZIMUTH ERROR* FOR A 100-YARD BASELINE

Angular Displacement	Approximate Equivalent in Linear Measure
$\frac{1}{60}°$	¾ inch
$\frac{1}{2}°$	2 feet
$1°$	4 feet

*Note that one minute of arc ($\frac{1}{60}°$) is equal in magnitude to the daily change of position of sun at horizon within approximately 2 days of the solstices. Thirty minutes of arc corresponds to the angular diameter of the sun or moon as well as to the approximate daily change of position of sun at horizon around the time of the equinoxes.

lateral shift in inches and feet for angles of one degree, $\frac{1}{2}°$, and one minute of arc, or $\frac{1}{60}°$, for the 100-yard baseline.

First, suppose you watch the sun's movement when it lies close to the equinoxes, when it moves the fastest. Table A.2 shows the daily shift in azimuth of sunrise (set) around these times of year. By how much (in inches) will the backsight be shifted from day to day? Next, suppose similar observations are made around the times of the solstices, when the sun moves very slowly. Again consult Table A.2 for the shift as you attempt to determine the daily amount by which you would expect to move the backsight. Could you devise a plan to circumvent the obvious difficulty in charting the day-to-day motion of the sun around the solstices? This is a fascinating experiment you can easily try out anywhere, including, for short baselines, on the roof of a building.

TABLE A.2. THE AZIMUTH OF SUNRISE FOR AN OBSERVER IN LATITUDE 20° N

Time	Daily Change in Azimuth in Minutes of Arc
Equinoxes	25.3
Equinoxes ± 1 day	25.2
Equinoxes ± 2 days	25.1
Solstices	0.0
Solstices ± 1 day	0.3
Solstices ± 2 days	1.2

CHAPTER FOUR
POWER FROM THE SKY

I. Here's how to make your own eclipse table. If the Eclipse Table in the Dresden Codex corresponds closely to the occurrence of eclipses in reality, then we should be able to create a table somewhat like it out of observed data taken over a reasonable time period. Table A.3 lists the dates of all eclipses, lunar and solar, total and partial, visible from Yucatán over the ten-year period 1997–2006. You can use it to construct a table similar to the Eclipse Table in the Dresden Codex. The last column tabulates the number of days between one eclipse and the next. First look at the lunar eclipses. Arrange them in chronological order and express the time difference between them, if possible, in terms of chains of 177-177-177 . . . 148 (Eclipse), just as we did for the Dresden Table (Table 4.1, page 113). (Remember that these numbers occasionally vary by ± I day in the table.) Second, do the same, considering only the solar eclipses. Finally, perform the same exercise for solar and lunar eclipses taken together.

Is your resulting table a reasonable approximation to what is found in the Dresden Eclipse Table for each of the three trials? Are there any significant differences between the format of the Dresden Table and the

TABLE A.3. ECLIPSES VISIBLE FROM YUCATÁN 1997–2006

			No. of Days Lapsed to Next Eclipse
1997	March 23	Lunar	340
1998	February 26	Solar	15
1998	March 13	Lunar	679
2000	January 20	Lunar	339
2000	December 25	Solar	354
2001	December 14	Solar	16
2002	November 20	Lunar	176
2003	May 15	Lunar	177
2003	November 8	Lunar	354
2004	October 27	Lunar	163
2005	April 8	Solar	16
2005	April 24	Lunar	324
2006	March 14	Lunar	

astronomical reality expressed in the data given in this exercise? Do you think the Eclipse Table in the Dresden Codex is an eclipse warning table? If so, does it warn of solar eclipses, lunar eclipses, or both?

2. The Venus Table in the Dresden Codex tabulates four subintervals in the cycle of Venus that add up to 584 days, but the real Venus synodic period is 583.92 days. Left unchecked, the written calendar would pull ahead of the real Venus by .08 day per Venus cycle. The problem of keeping time is similar to that of marking lunar synodic months by alternating 29- and 30-day cycles, which depends crucially on how accurately you can sight full or new moons. In this case you would be marking dates of heliacal risings of Venus. Suppose first that these events occur precisely 583.92 days apart. After how many years will the error accumulate to one full day?

But the problem is still more complicated! A correction page that comes with the Venus Table indicates that the Maya were well aware of the need to correct their ephemeris. This they did by dropping four days from the count on different occasions—the opposite of what we do when we add a day to our calendar year to make a leap year, thus allowing our year (of 365 days) to catch up to the real sun (365.2422 days). The choice of four and occasionally eight day corrections was made so that the Venus events would fall on particular name days in the 260-day

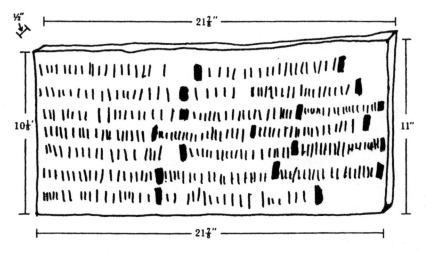

Figure A.3 A contemporary Maya "calendar board."

ritual day count. Suppose further that Venus heliacal rises and sets can be timed accurately to three days on the average. In this case, how many years would it take before the calendar shortfall would become notice-able?

3. Figure A.3 is a reproduction of a Chamula *Potalk'ak'al,* "Counter of Days," or solar calendar. Brought to light by anthropologist Gary Gossen in 1969,[1] it measures approximately 22 inches by 10 inches and records a calendar. Rendered in charcoal on wood, each mark stands for a day, the last mark in each cycle being most heavily emphasized. What cycles do you think are being marked out and how was the calendar keeper arranging them? What, if anything, about this contemporary cal-endar resonates with what we know about the ancient Maya calendar and counting system?

ARCHAEOASTRONOMICAL FIELDWORK

HOW TO DETERMINE THE AZIMUTH OF AN ALIGNMENT BY SUN FIX

The essential equipment for collecting alignment data in the field consists of a reliable surveyor's transit (theodolite) with altitude and azimuth scales, and an accurate watch. The watch may be set by time signals emanating from the radio station of the National Bureau of Standards, WWV (transmitting at 2.5, 5.0, 10.0, 15.0, and 20.0 MHz), or CHU of the Dominion Observatory (3.38 and 6.76 MHz). A copy of *The American Ephemeris and Nautical Almanac* for the year of observation is also useful. Computer programs for programmable calculators are available to reduce the data (see Appendix C: Archaeoastronomical Resources).

As an example of how typical field procedure is applied, we adopt a standard surveying technique for fixing the azimuth (the direction measured in degrees from true or astronomical north clockwise, that is, to the east) of a line, with special modifications and simplifications to accord with the archaeological environment. Figure B.1 illustrates how to determine the true azimuth of a wall. The observer places and levels the transit near a point *y* on the wall where the original structure still lies intact. He or she then measures off the perpendicular distance between *y* and the point where a plumb line from the transit strikes the ground (call it *x*).

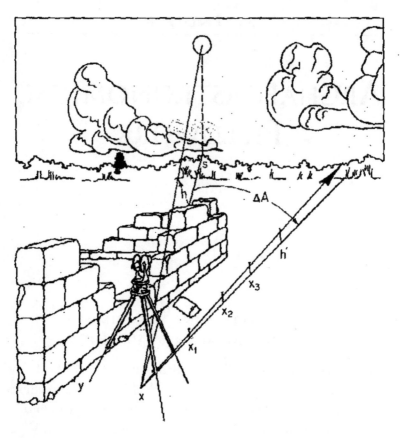

Figure B.1 How the archaeoastronomer measures alignments with a surveyor's transit.

This distance xy is marked at three or four places along the wall (x_1, x_2, x_3 ...). Then readings from the azimuth scale, which may be preset to zero, are recorded and averaged. Next, the telescope is raised to the level of the horizon and a reading on the altitude scale is taken. To recall the original situation correctly when later reductions are made, it is desirable to make a sketch showing transit placement and the general direction of both the wall and the sun.

To provide the correct information for altering the azimuths to true readings, one must observe a celestial body, usually the sun. If the transit is not equipped with a suitable solar filter for direct sighting, the solar image may be projected on a white card. Under no condition should an observer look at the sun directly without protection. Once the transit is directed at the sun and that object is positioned precisely in the middle of the field (sets of vertical and horizontal crosswires are a necessary accessory on all transits used in archaeoastronomical fieldwork), the observer calls "mark," the watch is read, and the azimuth xs and altitude h of the sun are recorded. If the sun's disk can be centered manually in the transit to within 0.5 minute of arc (⅟₆₀ of its own diameter), and a 2-second error is allowed on the watch, the method will be accurate to about 1 minute of arc (provided that the sun lies 10° or more above the horizon). While this is already more precise than the deviation from a straight line of any ancient wall that might be encountered in the field, for long-distance baselines it may be equal to errors incurred by naked-eye sighting.

A second method, though far less accurate, for reducing apparent to true azimuth may be used under conditions of poor solar visibility. This involves the use of the magnetic compass with which the transit is equipped, or a separate instrument such as a Brunton compass, which possesses a relatively precise sighting apparatus and a leveling device. After taking azimuth readings on the wall while keeping the transit fixed, allow the compass needle to swing and then simply record the deviation of the needle from the direction in which the instrument is already pointing. Because the magnetic compass does not actually align with true north, contemporary magnetic correction tables are needed[1] The magnetic alignment technique is not capable of yielding precision better than 1°. Compass needles are known to undergo long-term, annual, and even daily variations. Furthermore, local magnetic anomalies amounting to several degrees are not uncommon, therefore even a careful observer cannot expect precision from this technique.

Basically, one wishes to determine, via spherical trigonometry, the azimuth of the sun and then to compute the azimuth of the alignment measured relative to that of the sun:

Hour angle of sun (HA) in degrees
$$= (GMT - 12) \cdot 15 - LONG - EOT \cdot 15$$
Altitude of sun
$$= Arcsin\ (Sin(LAT) \cdot Sin(DEC) + Cos(LAT) \cdot Cos(DEC) \cdot Cos(HA))$$
Azimuth of sun $= Arcsin\ (Sin(HA) \cdot Cos(DEC) \div Cos(ALT))$
Azimuth of sun
$$= Arccos((Sin(DEC) - Sin(LAT) \cdot Sin(ALT)) \div (Cos(LAT) \cdot Cos(ALT)))$$

Where   • = multiplied by
       GMT = Greenwich Mean Time
        EOT = Equation of time
        LAT = Latitude of site
     LONG = Longitude of site
         HA = Hour angle of sun
       DEC = Declination of sun
        ALT = Altitude of sun

Note: The azimuth of the sun equals the azimuth from the Arccos if the azimuth from the Arcsin is negative. If the azimuth from the Arcsin is positive, then the azimuth of the sun equals 360 minus the azimuth from the Arccos.

APPENDIX C

ARCHAEOASTRONOMICAL RESOURCES

These references give basic information for computing positions of celestial bodies:

Bretagnon, P., and J-L. Simon. *Planetary Programs & Tables from −4000 to +2800.* Richmond: Willmann-Bell, 1986.

Duffet-Smith, P. *Practical Astronomy with Your Calculator.* Cambridge: Cambridge University Press, 1981.

Jones, A. *Mathematical Astronomy with a Pocket Calculator.* New York: Wiley, 1978.

Meeus, J. *Astronomical Formulae for Calculators.* Richmond: Willmann-Bell, 1982.

Montenbruck, O. *Practical Ephemeris Calculations.* New York: Springer-Verlag, 1989.

Montenbruck, O., and T. Pfleger. *Astronomy on the Personal Computer.* New York: Springer-Verlag, 1991.

Among the useful sources that give tabulations of celestial positions are:

Goldstine, H. "New and Full Moons, 1001 B.C. to A.D. 1651." Philadelphia: American Philosophical Society Memoirs, 94 (1973).

Meeus, J., C. Grosjean, and W. Vanderleen. *Canon of Solar Eclipses.* Oxford: Pergamon, 1966.

Meeus, J., and H. Mucke. *Canon of Lunar Eclipses −2002 to +2526.* Vienna: Astronomisches Buro, 1979.

Oppolzer, T. von. *Canon der Finsternisse* [1887]. New York: Dover, 1962.

Stahlman, W., and O. Gingerich. *Solar and Planetary Longitudes for Years −2500 to +2000 by Ten-Day Intervals.* Madison: University of Wisconsin Press, 1963.

Tuckerman, B. "Planetary, Lunar and Solar Positions 601 B.C. to A.D. 1, A.D. 2 to A.D. 1649." 2 vols. Philadelphia: American Philosophical Society Memoirs, 56 (1962) and 59 (1964).

Recommended sky simulation programs:

Voyager 1.2. The Interactive Desktop Computer (1988), Carina Software, San Leandro, CA. See also Voyager II, the Dynamic Sky Simulator, Version 2.0.

The Sky Astronomy Software for Windows, Version 2.0 (available from Astronomical Society of the Pacific, 390 Ashton Ave., San Francisco, CA 94112).

EZ Cosmos Version 3.0, Future Trends Software, 1601 Osprey Dr., Suite 102, Desoto, TX 75115

Planets and Moon and Sun, Charles Kluepfel, 11 George St., Bloomfield, NJ 07003

Free/shareware programs in the Info-Mac Library include:

Distant Suns	Starry Night
Sky Chart	Harpstars
Mac Astro	Micro Projects

For data on the deviation of magnetic from true north, see U.S. Geologic Survey, 1990, *The Magnetic Field of the Earth* (Maps). Denver: The Survey.

Finally, these are among the recent publications that deal with archaeoastronomical method and theory:

Aveni, A. "Archaeoastronomy," *Advances in Archaeological Method and Theory* 4:1–79 (1981).

Doggett, L., and B. Schaefer. "Lunar Crescent Visibility," *Icarus* 107:388–403 (1994).

Purrington, R. "Heliacal Rising & Setting, Quantitative Aspects," *Archaeoastronomy Supplement to the Journal for the History of Astronomy* 12:572–585 (1988).

Schaefer, B. "Predicting Heliacal Risings and Settings," *Sky & Telescope* 70(3):261–263 (1985).

———. "Heliacal Rise Phenomena," *Archaeoastronomy Supplement to the Journal for the History of Astronomy* 11: S19–34 (1987).

———. "Extinction Angles and Megaliths," *Sky & Telescope* 73(4):426–427 (1987).

———. "Refraction by Earth's Atmosphere," *Sky & Telescope* 77(3):311–313 (1989).

———. "Astronomy and the Limits of Vision," *Vistas in Astronomy* 36:311–361 (1993).

NOTES

CHAPTER ONE
INTRODUCTION

1. H. Frankfort, *The Intellectual Adventure of Ancient Man* (Chicago: University of Chicago Press, 1946), 136.
2. B. de Sahagun, *Florentine Codex. General History of the Things of New Spain Book 2. The Ceremonies*, ed. A. J. O. Anderson and C. E. Dibble (University of Utah School of American Research, 1981), p. 216.

CHAPTER TWO
THE NAKED SKY

1. The altitude or angle of elevation of the celestial pole above the horizon is equivalent to the observer's latitude. This relation can be useful in considering how the constellations are arranged in the sky as viewed from different latitudes. If we stand at the north geographic pole, the Pole Star will lie at the zenith or overhead position (altitude = latitude = 90° north). Imagine that we begin a journey from the north geographic pole to the equator, watching the Pole Star as we go. As we progress southward, Polaris will fall lower and lower in the northern sky (by one degree of altitude for each degree of latitude covered on the earth's surface). At New York (latitude 42° north), the altitude of the Pole Star will be 42°, or nearly halfway to the zenith. At Miami (latitude 26° north), we will see Polaris at an altitude of 26° above the north point of the horizon, and so on. Finally, when we arrive at the earth's equator (latitude 0°), Polaris will coincide with the north point of the horizon exactly. If we pass south of the equator, the Pole Star will dip below the horizon and the south celestial pole will rise to an altitude of 23° at Rio de Janeiro (latitude 23° south), 34° at Buenos Aires (latitude 34° south), and 78° at Mt. Erebus, Antarctica (latitude 78° south), etc.

2. Taking the argument to the extreme, an observer situated on the earth's equator (latitude 0°) will witness in the course of a year about twice as many stars as an observer at either of the geographic poles (latitude = 90° north or south). In the former case the celestial poles are situated exactly at the northern and southern horizon. All objects rise and set at 90° angles to the horizon and every constellation on the globe eventually can be seen. In marked contrast, a polar observer never sees more than half the objects in the sky; all of them pursue horizontal arcs around the celestial pole, which lies at the zenith or overhead point.

3. R. Frazer, ed., *The Poems of Hesiod* (Norman: University of Oklahoma Press, 1983), "Works & Days," lines 609–614.

4. The word *solstice* is indeed an appropriate term to describe the behavior of the sun at this time of the year—it means "sun stand" in Latin.

5. This is because these latitudes (23.5° north and south respectively) possess the same angular distance from the earth's equator as does the sun's northernmost and southernmost declination (its angular distance from the celestial equator).

6. While we call these the summer and winter solstice dates in the northern hemisphere, the terms should be reversed when we deal with the southern hemisphere latitudes. To lessen confusion and to avoid hemispheric bias, we will use the terms June or December solstice instead of the seasonal terms.

7. H. Livermore, ed., *Royal Commentaries of the Incas by Garcilaso de la Vega*, Vol. I (Austin: University of Texas Press, 1966), p. 116.

8. B. de Sahagún, *op. cit.* Chapter I, note 2, Book 7 (1953), p. 3.

9. While the sun moves 1° per day, the moon shifts 13° per day from west to east among the stars.

10. A. Heidel, ed., *The Babylonian Genesis (Enuma Elish)* (Chicago: University of Chicago Press, 1942), p. 45.

11. See B. Tedlock, *Time and the Highland Maya* (Albuquerque: University of New Mexico Press, 1982), pp. 190–197. Whereas twelve or thirteen lunar synodic months can be framed within a tropical year, the number of lunar sidereal months in a year can be thirteen or even fourteen. In fact, thirteen lunar sidereal months nearly approximate the length of twelve lunar synodic months:

$$13 \times 27.32166 \text{ days} = 355.1816 \text{ days}$$
$$12 \times 29.53059 \text{ days} = 354.3671 \text{ days}$$

This may be one reason why the Maya zodiac (p. 98 and Figures 2.15 and 4.10) consisted of thirteen, not twelve, constellations. Moreover, the description of the Inca *ceque* system of Cuzco given by the chroniclers offers a hint that the sidereal month and, in particular, a twelve-month period was an important part of the calendar. See the discussion beginning on p. 169.

12. A tiny wobble in the lunar orbit with a period of 173 days extends the range 9 arc minutes farther in either direction.

13. B. de Sahagún *op. cit.*, note 8, pp. 36–38.

14. This period is slightly less than a sidereal lunar month because the node itself regresses; that is, it moves among the stars in a direction opposite the motion of the moon. In effect, it meets the moon before the moon can complete a cycle with respect to the stars.

15. J. Pritchard, *Ancient Near Eastern Texts Relating to the Old Testament* (Princeton: Princeton University Press, 1955), p. 282.

16. B. van der Waerden, *Science Awakening* (Leiden: Noordhoff, 1974), p. 39.

17. Gary Urton, *At the Crossroads of the Earth and the Sky* (Austin: University of Texas Press, 1981), Ch. 6.

18. These dates vary noticeably with both the latitude of the observer and, over periods of hundreds of years, with the epoch of observations. See B. Schaefer (1986), "Atmospheric Extinction Effects on Stellar Alignments," *Archaeoastronomy Supplement to the Journal for the History of Astronomy*, No. 10 (JHA, xvii) pp. S32–S42 and my *Skywatchers* for technical details. Also, terms 3. and 4., in which sun and star are situated on opposite horizons, are often termed achronal rising and setting.

19. For details see D. Freidel, L. Schele, and J. Parker, *Maya Cosmos* (New York: Morrow, 1993).

20. See Gary Urton (1978), "Beasts and Geometry," *Anthropos* 73:32–40; (1978) "Orientation in Quechua and Incaic Astronomy," *Ethnology* 17(2):157–167. Also, Urton, *op. cit.*, note 17, p. 201.

21. Urton, *op. cit.*, note 17, pp. 193–196.

22. C. Levi-Strauss, *From Honey to Ashes* (New York: Harper, 1973), and G. Reichel-Dolmatoff, *The Shaman and the Jaguar* (Philadelphia: Temple University Press, 1975). See Urton's book (note 17) for a full list of references.

23. A common misconception is that ancient cultures measured celestial periodicities to several decimal places. This often leads to the notion that telescopes or other high-tech equipment was necessary. The Maya understanding of the lunar synodic month is quite similar, even though their mathematics employed neither fractions nor decimals. In fact, they utilized both 29- and 30-day synodic months, each represented by separate coefficients preceding a hieroglyph that stands for lunar phase. Portions of a Maya document of the preconquest period (pp. 51–58 of the Dresden Codex), discussed in detail in Chapter 3, express the precision in the length of the month by offering a list of months of both kinds; these are interposed in an irregular way, so that if one wishes to determine the phase of the moon for any date in the table, it always will agree with observation. Yet no mention of an exact lunar synodic period is made, nor is there any indication that the precise duration of the lunar cycle ever was calculated. So, date-reaching (not the computation of the lunar phase cycle to four decimal places) was the paramount goal of the Maya calendar system.

24. See, for example, A. Pannekoek, *A History of Astronomy* (New York: Dover, 1961), and my *Skywatchers of Ancient Mexico.*

25. In my *Empires of Time* (New York: Kodansha, 1993), Ch. 6, I show how our Gregorian calendar reform is mirrored in the tenth-century Maya reform of the Venus calendar.

26. Evidence provided by Spanish chroniclers is unclear on whether the Inca intercalated, for often these Spanish priests confused Inca astronomical concepts with those of Renaissance Europe. Most of them agreed that the Inca calendar employed twelve months, but they also tell us that the four most important solar ritual ceremonies followed the solstices and equinoxes and that these ceremonies also commenced at the observation of the first crescent moon following the appropriate equinox/solstice.

27. A full discussion of the conditions under which many exotic phenomena in the realm of the atmosphere optics can be found to occur can be found in M. Minnaert, *The*

Nature of Light and Color in the Open Air (New York: Dover, 1954). And an excellent presentation, with photographs, of the aurora borealis can be found in S.-I. Akasofu, "The Aurora Borealis," *Alaska Geographic Society* 6, No. 2 (1979).

CHAPTER THREE
STANDING STONES AND STARS

1. J. N. Lockyer, *The Dawn of Astronomy* (London: Cassell, 1894).
2. B. Somerville, "Orientation," *Antiquity* 1:33–34, 1927.
3. G. Hawkins, "Stonehenge: A Neolithic Computer," *Nature,* January 27, 1964.
4. *Diodorus of Sicily,* ed. H. Heineman, Book II, 47:1–48 (Cambridge: Harvard University Press, 1932).
5. R. J. C. Atkinson, "Moonshine on Stonehenge," *Antiquity* 40:212–216, 1966.
6. G. Hawkins, *Stonehenge Decoded* (New York: Dell, 1965), p. vii.
7. J. Hawkes, "God in the Machine," *Antiquity* 41:91–98, 1967.
8. A. Aveni, "The Thom Paradigm in the Americas," in *Records in Stone,* ed. C. Ruggles (Cambridge: Cambridge University Press, 1988), pp. 442–472.
9. See, for example, the texts edited by Aveni, Williamson, Ruggles, and Judge and Carlson cited in the General Bibliography.
10. R. Castleden, *The Stonehenge People* (London: Routledge, 1987), p. 29.
11. They would not be the only culture to do so. The ancient Maya, for example, styled many of their temples after their houses. At Uxmal's Nunnery, a frieze depicts a series of Maya huts that have the same basic form as the building on which they are carved.
12. Castleden, *op. cit.,* note 10, p. 200.
13. *Ibid.*
14. E. Fernie, "Stonehenge as Architecture," *Art History* 17(2):147–159, 1994.
15. Castleden, *op. cit.,* note 10.
16. Actually, the sun would have been positioned so that the viewer, standing at the center, could capture the sun's disk sitting on the horizon on the first day of summer in the Heel Stone *gateway,* when the sun reached its northern standstill about 3000 B.C. On the other hand, the first gleam of sunrise would have been viewed there a little over 10 days before and after the summer solstice, which would have provided an accurate way of fixing the solstice date. This could have been done by counting the number of days it would take to return back to the same place (about 20 days), halving the difference, and adding it on to the first observation.
17. The regression period is 6,797.1 days or 18.61 seasonal years. This equals 230.17 lunar synodic months (months of the phases), or 249.78 lunar draconic months (months measured by the interval between successive passages of the moon by a node of its orbit).
18. 6,940·days = 19.00 seasonal years, 235.01 lunar synodic months, and 255.03 lunar draconic months.
19. A. Thom, *Megalithic Lunar Observatories* (Oxford: Oxford University Press, 1971), p. 11.
20. *Ibid.,* p. 23.
21. See, for example, A. Thom, *Megalithic Sites in Britain* (Oxford: Oxford University Press, 1967), p. 21.

22. Hawkins, *op. cit.*, note 6, p. 97.
23. C. Ruggles, *Astronomy in Prehistoric Britain and Ireland* (New Haven: Yale University Press, 1996), and R. Clead, K. Walker, and R. Montague, *Stonehenge in its Landscape* (London: English Heritage, 1995).

CHAPTER FOUR
POWER FROM THE SKY

1. Stephens, *Incidents of Travel in Yucatan* (NY: Harper, 1843), Vol. 2, p. 307.
2. For a detailed discussion see K. Taube, *The Major Gods of Ancient Yucatan* (Washington: Dumbarton Oaks Research and Library Collection, 1992).
3. J. Torquemada, *Monarquia Indiana* (Mexico: Ed. Porrua, 1975) [1723], Vol. I, Book 2, Ch. 44, p. 188.
4. F. Tezozomoc, *Cronica Mexicana* [1548], ed. Orozco y Berra (Mexico, 1878), p. 574. I have identified St. Peter's keys to the golden gate with some of the stars in the Western zodiacal constellation of Aries, while *xonecuilli*, in my opinion, could be either our Little Dipper or the Southern Cross with Alpha and Beta Centauri added. The Fire Drill, used to carry New Fire into the household at the end of a long cycle (52 years) of the calendar, almost certainly was equated with the belt and sword of Orion. Some of these star groups are pictured in a manuscript (Figure 4.2) written by Father Bernardino de Sahagun, one of the more reliable chroniclers. (Of course, we cannot expect constellations to show up in these manuscripts as modern astronomers represent them, that is, one-to-one templates of the sky with the brighter stars represented by larger circles and the fainter stars by smaller ones.)
5. A. Tozzer, *Landa's Relación de las Cosas de Yucatán* (Cambridge, MA: Papers, Peabody Museum, 1941), p. 133.
6. J. de Acosta, *Historia Natural y Moral de las Indias* (Salamanca, 1589), Book VI, p. 6.
7. A. Tozzer, *op. cit.*, note 5, p. 169.
8. M. Coe, *Breaking the Maya Code* (London: Thames & Hudson, 1992), tells this fascinating story in all its detail.
9. There is evidence that the cycle may be connected to the human gestation period as well as to lunar eclipse and Venus appearance cycles.
10. Anthropologist Barbara Tedlock, in *Time and the Highland Maya* (Albuquerque: University of New Mexico Press, 1982) gives a detailed narration of this divination ritual, which is far more complex than I have indicated. See especially her Chapter 7.
11. J. E. S. Thompson, *A Commentary on the Dresden Codex* (Philadelphia: American Philosophical Society, 1972), p. 62.
12. One of the most thorough studies of the table is by H. Bricker and V. Bricker, "Classic Maya Prediction of Solar Eclipses," *Current Anthropology* 24(1):1–24, 1983.
13. Z. Nuttall, "The Periodical Adjustments of the Ancient Mexican Calendar," *American Anthropologist* (n.s.) 63:498, 1904.
14. Written 08. Note that the 0 is left in place over three dots and a bar.
15. See A. Aveni, "The Moon and the Venus Table: An example of Commensuration in the Maya Calendar" in *The Sky in Mayan Literature*, ed. A. Aveni (Oxford: Oxford University Press, 1992), pp. 87–101.

16. To state these relationships in more precise quantitative terms: **Venus Rounds:** 5 × 583.92 (5 Venus cycles) = 2,920 days − 0.40 day; **Seasonal Years:** 8 × 365.2422 (8 tropical years) = 2,920 days + 1.94 days; **Moon Phase Cycles:** 99 × 29.530589 (99 months of lunar phases) = 2,920 days + 3.53 days.

17. B. de Sahagun. *op. cit.,* Ch. 1, note 2, Book 10, pp. 191–192.

18. These iconographic observations were first collected together and analyzed by University of Illinois art historian Virginia Miller in "Star Warriors at Chichén Itzá," in *Word and Image in Maya Culture,* ed. W. Hanks and D. Rice (Salt Lake City: University of Utah Press, 1989), pp. 287–305.

19. F. Lounsbury, "Astronomical Knowledge and Its Uses at Bonampak, Mexico," in *Archaeoastronomy in the New World,* ed. A. Aveni (Cambridge: Cambridge University Press, 1982), pp. 143–168.

20. Alternatively they may stand generically for stars that make up these constellations.

21. Both art historian Ellen Baird, "Stars and War at Cacaxtla" in *Mesoamerica after the Decline of Teotihuacan,* ed. R. Diehl and J. Berlo (Washington, DC: Dumbarton Oaks, 1989), pp. 105–121, and astronomer/Mayanist John Carlson, "Venus Regulated Warfare and Ritual Sacrifice in Mesoamerica" (College Park: Center for Archaeoastronomy, Technical Publication No. 7. 1991), trace this connection.

22. See Aveni, A., S. Gibbs, and H. Hartung, "The Caracol Tower at Chichén Itzá: An Ancient Astronomical Observatory?" *Science* 188:977–985, 1975 (quote on p. 977).

23. *Ibid.*

24. J. Kowalski, "Uxmal: A Terminal Classic Maya Capital in Northern Yucatan," in *City States of the Maya: Art & Architecture,* ed. E. Benson (Denver: Rocky Mountain Institute for Pre-Columbian Studies, 1986), p. 154.

25. See Aveni, *op. cit.,* Ch. 2, note 24, pp. 273–276. Recently Ivan Sprajc, "The Venus-Maize-Rain Complex in the Mesoamerican World View," *Journal for the History of Astronomy* 24:17–71, 1993, has suggested that the fit between the alignment of the building and the place where Venus first actually appeared in the sky may not be as accurate as the match obtained by reversing the direction of the alignment, that is, by viewing Venus's northerly extreme, which would be visible after sunset looking from Cehtzuc to Uxmal. (He was first to identify the site with Cehtzuc. We had previously thought it to be Nohpat.) Moreover, Sprajc showed that the planet was more frequently and more prominently viewed during the evening as opposed to the morning sky in Late Classic times. Nonetheless, I am still persuaded that the alignment was eastward-looking because the Governor's Palace, in addition to possessing a large staging area for the enactment of Venus-related ritual, is also the building that contains all of the relevant Venus iconography. Still more recently the epigraphers Victoria and Harvey Bricker ("Astronomical References in the Throne Inscription of the Palace of the Governor at Uxmal," *Cambridge Archeological Journal* 6:191–229, 1996) have identified a band of zodiacal signs running over the doorway of the Governor's Palace.

26. Closs, Aveni, and Crowley, "The Planet Venus and Temple 22 of Copan," *Indiana* 9:221–247, 1984, discuss the Chac-Venus association. We offer more examples of Maya buildings at other sites that were likely oriented to anticipate the disappearance and reappearance of the bright planet. In fact, there is only a single example outside Uxmal where the Venus glyph is exhibited as part of the design; interestingly, these

occur on masks on the north side of the Nunnery at Chichén Itzá (the facade facing the Caracol). Like the masks on the Palace of the Governor, they are arranged in groups of five, in agreement with the format of one Venus cycle per page of the Dresden Table.

27. J. Rivard, "A Hierophany at Chichén Itzá," *Katunob* 7 (3):51–55, 1971.

CHAPTER FIVE
CITY AND COSMOS

1. Garcilaso de la Vega, *Royal Commentaries of the Incas*, ed. H. Livermore (Austin: University of Texas Press, 1966), Vol: 2, p. 114.

2. P. Cieza de Leon, *La Crónica del Peru* [1550], ed. J. Munoz, et al. (Lima: Biblioteca Peruana, 1973).

3. For a thorough description and a map of the building, see J. H. Rowe, "An Introduction to the Archaeology of Cuzco," *Papers Peabody Mus., Harvard*, vol. 27 (No. 2) 1944.

4. J. Hemming, *The Conquest of the Incas* (NY: Harcourt Brace, 1970), p. 64.

5. Garcilaso de la Vega, *op. cit.*, note 1, Vol. I, p. 182.

6. Hunahpu and Xbalanque, the Maya creator twins of the *Popol Vuh* (ed. D. Tedlock, New York: Simon & Schuster, 1985), represent the sun and Venus while Apsu and Tiamat are the fresh and salt waters that meet at the shoreline of the Persian Gulf in the Babylonian Enuma Elish (ed. A. Heidel, Chicago: University of Chicago Press, 1942, Ch. 2, note 10).

7. J. Murra, "El 'Control Vertical' de un Máximo de Pisos Ecológicos en la Economía de las Sociedades Andinas," in *Visita de la Provincia de Léon de Huanuco [1562] Iñigo Ortiz de Zúñiga Visitador* (Huanuco, Peru: Universidad Nacional Hemilio Valdizán, 1972), pp. 429–476.

8. G. Urton, *op. cit.*, Ch. 2, note 17, especially Chapter 3.

9. C. Morris and D. Thompson, *Huanuco Pampa: An Inca City and Its Hinterland* (London: Thames and Hudson, 1985), have detected rudiments of *suyu-ceque* structures at Huanuco Pampa and the title of John Hyslop's *Inkawasi, the New Cuzco* (Oxford: British Archaeological Reports, IS 235, 1984) speaks for itself.

10. B. Cobo, *Historia del Nuevo Mundo* [1653] (Madrid: Biblioteca de Autores Españoles v. 91–92, 1956). All translations quoted here are by J. Rowe, "An Account of the Shrines of Ancient Cuzco," *Ñawpa Pacha* 17:27, 1979.

11. B. Cobo, *op. cit.*, note 10. Chronicler Cobo's hierarchical description of the *ceques* is reminiscent of Andean *quipus*. The *quipu* is a form of language—rather like braille. Typically it consisted of a thick cotton cord (the primary cord) from which were suspended thinner subsidiary cords. These often were of different colors, thus lending a visual element to the system. Each cord contained clusters of knots according to a decimal system. Hierarchically, the cords may be likened to *ceques*, and the knots to *huacas*. According to the chronicler Matienzo, the original description of the *ceque* system was transcribed from a *quipu*, primarily a device for record-keeping. The importance of the *quipu* is well attested to in the chronicles; for example, Guaman Poma discusses and illustrates the duties of *quipu-kamayoc*, or *quipu* specialists, and Garcilaso de la Vega's lengthy writings on the subject leave no doubt that one of the primary functions of the *quipu* was accounting.

12. See, for example, G. Urton, *The History of a Myth, Pacariqtambo and the Origin of the Incas* (Austin: University of Texas Press, 1990).

13. F. Guaman Poma de Ayala, *El Primer Nueva Corónica y Buen Gobierno* [1584–1614], ed. J. Murra and R. Adorno (Mexico: Siglo XXI, 1980), ff. 353–354, 883–884.

14. Cf. B. Bauer and D. Dearborn, *Astronomy & Empire: In the Ancient Andes* (Austin: University of Texas Press, 1995).

15. (Anonymous Chronicler) V. Maurtua (ed.) "Discurso de la Sucesión i Gobierno de los Yngas," in *Juicio de Limites Entre el Perú y Bolivia.* Prueba Peruana 8:149–165, (Madrid: Chunchos, 1906).

16. Bauer's (*op. cit.*, note 14, p. 94) delineation of the relevant and relatively crooked *ceques* in this region does not support the idea that a sight line of any significant precision pointed to the sunrise on the day of its passage through the zenith. But there may be more at issue in judging Inca astronomy than precision alone, as I shall argue below.

17. C. de Molina, *Relacion de las Fabulas y Ritos de los Incas* [1573] (Lima: D. Miranda, 1943), p. 25.

18. But for the addition of the stars to the picture, the whole scheme of date-finding is reminiscent of what we found at Stonehenge.

19. See R. T. Zuidema, "Catachillay: The Role of the Pleiades and of the Southern Cross and Alpha and Beta Centauri in the Calendar of the Incas," in *Ethnoastronomy and Archaeoastronomy in the American Tropics,* ed. A. Aveni and G. Urton (New York: Transactions of the New York Academy of Sciences, 1982), Vol. 385. The Huarochiri manuscript, the only native document with links to the precontact period, says of the Pleiades: "next are the ones we call the Pleiades; if they come out at their biggest people say, 'This year we'll have plenty.' But if they come out at their smallest, people say, 'We're in for a very hard time.' " F. Salomon and G. Urioste (ed.), *The Huarochiri Manuscript* (Austin: University of Texas Press, 1991), p. 133.

20. Using astronomical horizon observations to determine planting dates are very similar to what Urton (*op. cit.*, note 8) reports in his study of the present-day village of Misminay, northwest of Cuzco. Also the Inca utilized an 8-day week, so that a 41-fold division of the year existed as well ($8 \times 41 = 328$).

21. Cobo, *op. cit.*, note 11, pp. 172 and 185 respectively.

22. *Ibid.*, p. 185.

23. Bauer and Dearborn, *op. cit.*, note 14, pp. 76–80. They also have proposed other possible locations for these pillars.

24. *Ibid.*, pp. 81–89.

25. P. Cieza de Leon, *op. cit.*, note 2, p. 214. P. Sarmiento de Gamboa, *Historia de los Incas* [1572] (Buenos Aires: Emece, 1942), p. 175, after describing the horizon pillars, continues with this curious statement: "...he [Inca Tupac Yupanqui] put columns of stone in place and ordered the ground paved. On the flagstone he laid out graded rays to correspond to the direction of the sun as it entered through holes (in the columns), so that the whole device functioned as an annual clock and there were certain people who kept the count on these clocks and accordingly notified the pueblo of the different times for planting and harvesting." Garcilaso (*op. cit.*, note 1, vol. I, p. 117) gives a similar description.

26. Cobo also refers to alignments to the rising positions of the sun at the solstices taken to the east. Here the distant mountains served as the markers.

27. Jorge Luis Borges and Adolfo Bioy Casares, "On Exactitude in Science," in *Fantastic Tales* (New York: Herder and Herder, 1971), p. 23.

CHAPTER SIX
THE WEST VS. THE REST?

1. A. Aaboe, "Scientific Astronomy in Antiquity," in *In The Place of Astronomy in the Ancient World.* (London: Philosophical Transactions of the Royal Society, 1974), p. 21.
2. D. Schmandt-Besserat, *Before Writing*, 2 vols. (Austin: University of Texas Press, 1992).
3. R. Thompson, *Reports of Magicians and Astrologers of Nineveh and Babylon in the British Museum* (London: Luzac, 1900), Vol. 2 (91 and 81).
4. P. Huber, "Early Cuneiform Evidence of Venus," in *Scientists Confront Velikovsky*, D. Goldsmith (ed.) (Ithaca, NY: Cornell University Press, 1977).
5. B. van der Waerden, *Science Awakening, II, The Birth of Astronomy* (Leyden: Noordhoff, 1974), p. 39.
6. R. Thompson, *op. cit.*, note 3, 2(70).
7. The theme of the city as a reflection of the cosmos is found in other societies as diverse as the Maya and Aztecs of Mexico (see A. Aveni, "The Role of Astronomical Orientation in the Delineation of World View," in *The Imagination of Matter*, ed. D. Carrasco (Oxford: British Archaeological Reports IS 515), pp. 85–102 and the ancient Chinese, e.g., P. Wheatley, *Pivot of the Four Quarters* (Chicago: Aldine, 1971). Geometry literally means "land measure," and it is from the practical art of surveying and city planning that the Greeks took to applying this earth-bound principle to charting out the heavens.
8. Historian of mathematics Alan Bishop, "Western Mathematics: The Secret Weapon of Cultural Imperialism," *Race and Class* 32(2), 1990, has pointed out that there are other conceptions of space than the one held in the west. The Navajos, for example, neither subdivide nor objectify space. For them all events are described in terms of continuous motion; and in Papua, New Guinea, space is defined in terms of boundaries or edges rather than points and centers.
9. Ptolemy [Claudius Ptolemaios], *Syntaxis Mathematica*, ed. J. L. Heiberg (Leipzing Teubner, 1898), Vol. I, preface.

APPENDIX A
THINGS TO THINK ABOUT

1. G. Gossen, "A Chamula Solar Calendar Board from Chiapas, Mexico," in *Mesoamerican Archaeology, New Approaches*, ed. N. Hammond (Austin: University of Texas Press, 1974), pp. 217–253.

APPENDIX B
ARCHAEOASTRONOMICAL FIELDWORK

1. See, for example, J. Nelson et al., *Magnetism of the Earth*, U.S. Department of Commerce Publication 40–1 (Washington, DC: U.S. Department of Commerce, 1962).

ILLUSTRATION CREDITS

Grateful acknowledgment is made to the following for permission to reproduce material used in creating the figures in this book. Every reasonable effort has been made to contact the copyright holders of material used here. Omissions brought to our attention will be corrected in future editions.

Figures 2.1, 2.2, 2.3, 2.4, 2.5, 2.6, 2.7, 2.9a, 2.10, 2.11, 2.12, 2.13, 2.14, 4.6, 4.11b, 5.3, 5.7a, 5.8b, 5.9, 6.2, a, b, c, A.1 drawn by Ellen Walker.

2.8a, b, & c	Courtesy of Horst Hartung.
2.9b	Photograph by author.
2.15a	Courtesy of E. C. Krupp, Griffith Observatory.
2.15b	Courtesy of Brian Sullivan.
2.15c	Drawing by Ellen Walker; adapted from G. Urton, *At the Crossroads of the Earth and the Sky*, Austin, University of Texas Press, 1981.
3.1	From Inigo Jones, *The Most Notable Antiquity of Great Britain Vulgarly Called Stonehenge*, 1655. Courtesy of the Glasgow University Library, Special Collections Department.
3.2	Adapted from Anthony Aveni, *Ancient Astronomers*, Smithsonian, Washington, D.C., 1993.
3.3	Adapted from *Nature*, 200:306, 26 October 1963. Courtesy of Macmillan Magazines Ltd.
3.4	Courtesy of Margaret R. Curtis.
3.5a, c	Courtesy of Horst Hartung.
3.5b, d	Photos by author.
3.6a, b, c	Reprinted from R. Castleden, *The Making of Stonehenge*, Routledge, Hampshire, UK, 1993. Reprinted with the permission of Routledge.
3.7	Reprinted from R. Castleden, *The Stonehenge People*, Routledge, Hampshire, UK, 1987. Reprinted with the permission of Routledge.
4.1	Drawing by George E. Stuart.
4.2	Copyright © Patrimonio Nacional, Madrid.
4.3d	Courtesy of the Bodleian Library, Oxford; MS. Mex. d. l. Fol. 4r detail.
4.3e, f	Courtesy of the Bibliothèque National de France, Paris.
4.4	From *Codex Borgia*. Restoration by G. Diaz and A. Rodgers, Dover, New York, 1993. Used by permission of the authors.
4.5	Drawing by Peter Dunham, *Toward a New Definition of Literature*, M. Wrolstad & D. Fisher (eds.), Prager, New York, 1983, pp. 262–263. Used with permission.

4.7, 4.8   Reprinted from J. Villacorta and A. Villacorta, *Codices Mayas Reproducidos y Desarrollados,* Tipografia Nacional, Guatemala City, 1976.

4.9   From Anthony Aveni, *Conversing with the Planets,* New York, Times Books, 1992.

4.10a   Courtesy of E. C. Krupp, Griffith Observatory.

4.10b   Reprinted with the permission of the Peabody Museum of Archeology and Ethnology, Harvard University.

4.10c   Courtesy of Augusto Molina Montes.

4.11a   Photo by author.

4.12a   Reprinted from *The Ancient Maya, Fifth Edition,* by Robert J. Sharer, with the permission of the publishers, Stanford University Press. © 1946, 1947, 1956, 1983, 1994 by the Board of Trustees of the Leland Stanford Junior University.

4.12b   Photo by author.

4.13   Drawing by Anne Dowd. Courtesy of the Peabody Museum of Archeology and Ethnology, Harvard University.

p. 146   Photograph by George T. Keene. Reprinted with the permission of George T. Keene.

5.1   Photograph by Susan A. Niles. Reproduced with the permission of Susan A. Niles.

5.2   *Relacion de Antiquedades deste Regno del Piru.* Coll. Hist. De Peru, Lima 1951 [1613].

5.4   Courtesy of Craig Morris, American Museum of Natural History.

5.5   *El Primer Nueva Corónica y Buen Gobierno por Felipe Guaman Poma de Ayala.* [1615] ed. A. Posnansky La Paz, Inst. Tihuanacu de Antr., Etn. y Prehist, 1944.

5.6   Drawing by Brian S. Bauer. Reproduced with the permission of Brian S. Bauer.

5.7b   Photo by author.

5.8a   Adapted from R. T. Zuidema, *Ann. N. Y. Acad. Sci,* Vol. 385, 1982.

5.10   Reprinted from A. Aveni, "Tropical Archaeoastronomy," *Science, 213,* 161–171, Figure 8. Copyright 1981 American Association for the Advancement of Science. Reprinted with permission.

6.1   Courtesy of the Bibliothèque National de France, Paris.

A.2   Reprinted from A. Aveni, *Ancient Astronomers,* Smithsonian, Washington, D.C., 1993.

A.3   Drawing by Gary Gossen. Reproduced with permission of Gary Gossen.

B.1   From A. Aveni, Archaeoastronomy, *Advances in Archaeological Method and Theory, Vol. 4,* Figure 1.7. © 1981, New York Academic Press. Used with permission.

INDEX

Aaboe, Asger, 178
agriculture
  astronomy, 8
  calendar, 25–27, 53, 163
  crop predictions, 46
  grape harvest timing, 47
Alta Vista (Chalchihuites), 23–24
Ammizaduga, King, 182
Anaxagoras, 187–188
antizenith passage. *see* sun
Apollo, 3
archaeoastronomy
  alignment, methods used, 172–174
  development, 5
  interdisciplinary character, 76
architecture
  Inca
    cosmological expression, 154
    testimony to timekeeping, 55
  Maya. *see* Maya
  Pleiades and architectural alignment of,
    166, *166*, 167
  Stonehenge. *see* Stonehenge
Arcturus, 47
Ashurbanipal, King of Nineveh, 179
Assyrians eclipse prediction *(saros)*, 36–37,
  111
astrology
  Babylon, 185
  Greece, 191–192
  Maya, 6, 185
  Western, 48
astronomical cycles and life cycles, 52
astronomy
  agriculture, 8
  Aztec, 17
  Chinese, 17
  Inca
    agriculture, 8
    alignments, *172*
    ancestor worship, 8
    historical record, 161

methodology, 160
purpose, 160
unwritten, 174
Maya
  generally, 122
  religion linkage, 145
Mesoamerica, 98
positional, mathematics and
  record-keeping, 4
Atkinson, Richard, 68–70, 90
Aubrey, John, 175
Aztec
  astronomy, 17
  Hill of the Star, 17
  lunar cycle, 28
  militaristic cosmology, 128
  native constellation recognition, 99,
    100–101
  rabbit in the moon, 28, *29*
  solar eclipse, effect of, 34
  Tenochtitlan, 17, 98, 128

Babylon
  Aaboe, Asger, 178
  Ammizaduga, King, 182
  astrology, 185
  cuneiform records, 178–183. *see also*
    Venus Tablet of Ammizaduga
  speculated origin, 178
  forecasting, spatial and temporal, 181
  lunar eclipse records, 181
  Mars observations, 181
  planetary motion, approach to, 178
  Schmandt-Besserat, Denise, 179
  skywatching, 182
  Venus observations, 38, 40
    Maya (Dresden Codex), compared,
      183, 183–184
Banaba Island, 15
Bauer, Brian, 160, 173
Beru Island, 15
Big Dipper, 48

223